BRIGHT PARADISE

VICTORIAN SCIENTIFIC TRAVELLERS

By the same author

Fair Ophelia:
A Life of Harriet Smithson Berlioz

Oscar Wilde

Samuel Butler

BRIGHT PARADISE
VICTORIAN SCIENTIFIC TRAVELLERS

PETER RABY

Princeton University Press
Princeton, New Jersey

First published in the United States of America in 1997 by
Princeton University Press, 41 William Street, Princeton, New Jersey 08540

First published in Great Britain in 1996 by
Chatto & Windus Limited
Random House, 20 Vauxhall Bridge Road,
London SW1V 2SA

Random House Australia (Pty) Limited
20 Alfred Street, Milsons Point, Sydney
New South Wales 2061, Australia

Random House New Zealand Limited
18 Poland Road, Glenfield
Auckland 10, New Zealand

Random House South Africa (Pty) Limited
PO Box 337, Bergvlei, South Africa

Random House UK Limited Reg. No. 954009

ISBN 0–691–04843–6

Set by SX Composing DTP, Rayleigh, Essex
Printed and bound in Great Britain

Princeton University Press books are printed on acid-free paper and meet
the guidelines for permanence and durability of the Committee on
Production Guidelines for Book Longevity of the Council
on Library Resources

1 3 5 7 9 10 8 6 4 2

(Pbk.)

For Hamish

Contents

The World: Atlas

Valker's Geography, 1799

Preface and Acknowledgements

In writing a book which ranges broadly, I have been extremely conscious of the large gaps in my knowledge, and so more than usually dependent on the commentaries and detailed studies of specialists. The books I have found most helpful are mentioned in the bibliographical notes to each chapter, at the end of the book.

I would like to thank many friends for their generosity in conversations, in making suggestions, in recommending and often lending me books and material. I am especially grateful to Ted Chapman, Don Cupitt, Walter Henderson, Stephen Hugh-Jones, Murray Pollinger, Christopher Roper and Nick Wykes, to Diana and Patrick Ridsdill Smith for lending us their cottage in the Lot, where the better ideas seem to germinate in warmth and peace, to Beth Humphries for editorial assistance, and to Jenny Uglow for her patience and encouragement. David Hanke kindly read a draft, and offered invaluable advice: the mistakes and idiosyncrasies which remain are mine. My wife Elizabeth has had to accompany me on some exhausting mental travels, and, more voluntarily, on enjoyable journeys to Down, Ermenonville and Walden Pond. I would not have completed this book without her help.

For assistance in preparing the typescript and illustrations, I am indebted to Alan Russell, Barrie Smith and Bob Lashmar. I am grateful to the Principal and Trustees of Homerton College, Cambridge, for research support. I also thank the Librarian and staff of Cambridge University Library for their courtesy and assistance, and for permission to quote from manuscript material.

Gintrac, Lot – Swaffham Bulbeck, Cambridge

Introduction: To the World's Beginning

THE IDEA OF THE TRAVELLER HAS been elevated into a late twentieth-century industry, with its own infrastructure of guidebooks, hostels, eating houses and travel agents. The native inhabitants of the uttermost parts of the earth have an uneasy relationship with these nomadic peoples, welcoming them for their contribution to the local economy, resenting them for the mess of all sorts that most of them leave behind. Within the travellers' community, subtle variations and sub-species are recognised, ranging from soft to hard, from those with the security of round the world Trailfinders tickets to wiry soloists on battered bicycles, from packs or pairs to the resolute, obsessive individual. If you are really unlucky, you can even be accompanied to a desert island by a BBC television crew with logistical back-up. The whole world is suddenly available, and travel – or, in certain circles, the year out or the 'gap' year – has become not just a rite of passage but a way of life. Nowhere is safe: the Annapurna trail is as well trodden as the Pennine Way. A handful of countries are ignored, because a political regime or civil war make them expensive or dangerous to visit: Bhutan, Burma, Iraq. Some Muslim countries are spared, because their religious code imposes too severe a discipline. A few areas of South America and central Africa remain largely inviolate, because of the physical difficulties, or the incidence of disease; but the Indian subcontinent, South-east Asia, the Far East, north, east and south Africa and most of South America have been made available, many to the package tourist, but all to the backpacker or the 'adventure' tourist. Alternative holidays, adventure holidays, special interest holidays, safari holidays – a walk in the Venezuelan rain forest, ballooning in the Himalayas, motorcycling through India: you can choose between a 'luxurious remote lodge' or take up the challenge hinted at by an advertisement headed 'Mad dogs and Englishmen'. Travel has lost its strangeness, though not its allure.

The things we long to see, too, are no longer strange. They have been photographed and filmed, possessed already at second hand. We know, or think we know, what the forest peoples of the

Amazon basin look like; their representatives have flown out of jungle clearings to take part in discussions about the future of the rain forest. The Indonesian government has opened up Irian Jaya and encouraged, or at least allowed, a veteran traveller to ask a tribesman what human flesh tasted like. Naturalists have crawled through the highland vegetation and squatted in the same nest as the mountain gorilla. Images of the other have become disarmingly familiar: the tropical forest, the 'primitive' or 'savage', the rare species. Yet they still exert a powerful attraction. The very juxtaposition of the two opposing worlds and cultures disturbs, and the suddenness with which we can be transported from one time zone or climate to another both thrills and shocks. Our growing sense of the fragility of the environment, of the world itself, gives a new edge to the sense of discovery and wonder which still accompanies the most modest expedition. Underlying each new contact, direct or vicarious, is a sense of guilt and responsibility for the changes which history has brought about and to which present pressures are adding.

In *Mansfield Park*, Jane Austen has Lady Bertram bid William Price to go, for her convenience, to India, 'that I may have a shawl, I think I will have two shawls'. At the end of the imperial century, Wilde's Mrs Allonby, in *A Woman of No Importance,* encourages Gerald Arbuthnot to go to India with Lord Illingworth; she has less predictable, more sophisticated, tastes: 'Mind you bring me back something nice from your travels – not an Indian shawl – on no account an Indian shawl.' For the English in the nineteenth century, abroad, and especially the Empire and the colonies, existed to bring things back from.

The most vivid memories from my boarding-school in the late 1940s are images of empire. We slept in dormitories named Wellington, Gordon, Kitchener, Livingstone. On the shelves of the school library were rows of novels by G.A. Henty: I cannot remember ever reading one, but the names, or the rhythm of the titles, are etched on my mind as boldly as the illustrations on the tight cloth covers. Each Armistice Day, there was an impressive service at which the names of all the old boys who had died in the two world wars were read out, with a pause after each one. It was a long and moving list; the first lines of Rupert Brooke's poem were inscribed on the War Memorial, and, I think, read out at the service:

> If I should die, think only this of me:
> That there's some corner of a foreign field
> That is for ever England.

In my memory, at least, we sang the hymn 'I vow to thee, my country, all earthly things above . . .' Since Latin was our second language, a language of empire and conquest in which Caesar was always scattering the Gauls and the Belgae, we also knew what *Dulce et decorum est pro patria mori* meant: Wilfred Owen came later, with the disillusionment and reality of adolescence. We learned Latin by rote from Kennedy's Latin Primer, devised by the legendary headmaster of Shrewsbury in the 1840s, A.L. Kennedy, the oppressor of Samuel Butler. It was drummed into us, with the aid of terror, shame and the cane, by a benign-looking, white-haired, wholly terrifying Anglican clergyman of a headmaster. We were drilled and instructed in PE and swimming by an ex-army sergeant, who also appeared disconcertingly with a clipboard outside the latrines, to register our morning visits on his list. It was an education entirely suitable for an officer class; and somehow the idea of dying abroad for your country's good became bound up in my mind with the fictional exploits of Henty's heroes, and the stories of Gordon and Kitchener: *The Dash for Khartoum, With Kitchener in the Soudan.* The rest of the world, so remote from the closed, rigid routine of our rhododendron-enclosed cantonment in the Surrey hills, was a place to confront and conquer, to convert and civilise. There were no scientists among the rollcall of our household gods. Darwin did not feature in the curriculum. We were still children of the Empire, even as it was being unravelled.

For an early nineteenth-century child, the image of the world, from a comfortable vantage point inside Britain, was a complex and highly selective amalgam of allure and threat. There was Europe and, closest to home, France – the home of liberty, but also, paradoxically, of violent revolution, a set of terrors personalised in the figure of Napoleon. To the west lay the ex-colonies of the United States, whose ability to challenge and finally overtake Britain as a world power would grow through the century. To its north, Canada, less hospitable but more amenable, and to the south the vast but untapped countries of South America, Argentina and Brazil. Offshore were the plantations of the West Indies, an asset which would become more and more problematic as the slow and painful retreat from slavery took its course. To the east lay India, the great generator of wealth, which would be transformed during the century from a domain of the East India Company to the actual heart of the imperial theme, as indirect commercial control was converted into triumphant rule under the Queen Empress. Beyond, in the southern hemisphere, lay the penal colony of Australia, which would slowly change to a land of opportunity rather than punishment, and be followed by the new settlement of New Zealand, as a place where the unemployed, the frustrated and the adventurous could create a new way of life. Much closer to home, but somehow more mysterious and forbidding, was the continent of Africa, the secrets of its heartland protected by climate, terrain and disease. Other areas and countries, remoter, less accessible, attracted individual imaginations; and, as the century progressed, the impulse to penetrate every river, cross each desert, push through the ice-fields, became steadily more emphatic.

Until the eighteenth century, most expeditions and voyages had been undertaken from a mixture of motives. Perhaps the first truly scientific journey was the dual French expedition of 1735, one part aimed towards Lapland, one bound for the equator, to test whether the earth was fatter round its middle (Newton's theory, English) or at the diameter passing through the poles, as Cassini the French Astronomer Royal had suggested. The Lapland expedition had the easier time. It took La Condamine, headed for Quito, ten years to get back to Paris, and sixteen years passed before he published his report: *Journal du voyage fait par ordre du roi à l'équateur, servant d'introduction historique à la mesure des trois premiers degrés du méridien.* The controversy was by then of little interest, but La Condamine's epic journey, like Cook's voyages, served as a model for future scientific travellers, notably Humboldt. Even more inspirational, perhaps, were the stories of Jean Godin, a surveyor on the La Condamine expedition, and his Peruvian wife, Isabela. Jean Godin set off down the Amazon, leaving his pregnant wife in Peru, by way of a preliminary reconnaissance. It was twenty years before a message arrived to say that a boat was waiting to take Isabela to Para, and then on to Cayenne, where her husband survived in patient and rather passive expectation. The first part of the journey, through the eastern valleys of the Andes, was the most difficult. She set off with a party of forty-one, including two of her younger brothers and her nephew, a boy of twelve. They suffered every kind of disaster and misfortune: smallpox, fever, capsizing, desertion, separation, rape. Finally, Isabela found herself alone, and just alive. She stumbled off into the forest. When, nine days later, she met a party of Indians, she was semi-conscious and was wearing only 'the soles of the shoes of her dead brothers'. The Indians clothed her and took her to a mission station, from where she was canoed to a rendezvous with the waiting ship. The rumour of her terrible death had already reached her husband, and indeed France. Now it could be revised into a story of miraculous preservation: an Eve, returning from the Garden. Her amazing and protracted ordeal, *en route* from Peru to the France of the Enlightenment by way of the tropical forest of the Amazon basin, was a kind of human validation of the scientific purpose.

Most of the scientific voyages which preceded the great age of Victorian exploration were officially sponsored and financed: this was science in the service of the state. Captain Cook was sent, originally, to chart the course of the planet Venus from the vantage point of Tahiti. These eighteenth-century voyages were Royal Naval expeditions. If the world's oceans were charted, the navy could control them, and the knowledge of depths and tides could then be placed in the hands of the next instrument of power, the British merchant. The process of mapping and surveying is one of the most authoritative actions in the exercise of government. Brian Friel's play of 1980, *Translations*, movingly recreates the devastating ordnance survey of Ireland in 1833, in which the invading coloniser measures and renames the annexed territory. Friel creates perspective on this process through the character of Lieutenant Yolland, a reluctant map-maker, who recognises in his commanding officer the drive and energy of his own father, an engineer who builds roads all over the Empire: he was born in 1789, the very day the Bastille fell. 'I've often thought maybe that gave his whole life its character. . . . He inherited a new world the day he was born – The Year One. Ancient time was at an end. The world had cast off its old skin. There were no longer any frontiers to man's potential.' So the British marched across frontiers, mapping and naming as they went. Kipling's Kim, at the end of the nineteenth century, was educated to become a chain-man, a recruit for the great survey of India and a small part of the Great Game, pacing the streets of the remote, walled city of Bikaneer and calculating the distances by means of a rosary.

The great voyages of Captain James Cook, and his French counterpart Bougainville, provided the framework for the explorers of succeeding generations. Naval power, science and empire converged with superb economy. Cook effectively made Australia and New Zealand available for British colonisation, even if the choice of Australia as a penal colony was a case of *faute de mieux*; he disposed of the myth of an unknown southern continent. On board the *Endeavour* on Cook's first voyage in 1768 was the naturalist Joseph Banks, who would become president for forty-two years of the Royal Society, the great shaper of Kew Gardens and the patron and inspiration of generations of naturalists: his fellow naturalist was a pupil of Linnaeus, Douglas

Solander. The images of the south seas, or at least Tahiti, as a potential paradise were not altogether dispelled by Cook's death in the Sandwich Islands, on his way back on his third voyage from an attempt to find the Pacific outlet of the North-West Passage. In the Atlas to *Walker's Geography* of 1799, Cook's routes are clearly marked, including his voyage of 1773–75: 'Former tracks of J. Cook shewing the non Existance of a Terra Australis or Southern Continent'; and another inscription testifies how much the new map of the world stemmed from him: 'Owhehee where J. Cook was killed'. Everywhere they went the explorers left evidence of their presence like animals depositing scent traces on their territory: New Hebrides, New Caledonia, New South Wales; Drake's Harbour, Banks's Island. As a crowning tribute to the wealth of unknown plants in the new world, Botany Bay was named for the new science, with Point Solander and Cape Banks at either extremity.

The impulses which propelled the nineteenth-century long-distance travellers were diverse; most of those who ventured beyond Europe had, ostensibly, a definite objective: they set off on naval or military duty; for trade; to emigrate; to convert. (Many English women travelled to India to marry; but that can be seen as part of a tribal movement; they were not stopped at Immigration.) There was also a significant group who went to solve questions about the unexplored parts of the earth. Where was the source of the Nile? Did the Niger flow eastwards or westwards, and where if at all did it enter the sea? What lay in the centre of Australia? Was there a North-West Passage? These explorers, armed with clinometers and theodolites, formed the last wave of the mappers and charters.

Support for the sciences in the first half of the nineteenth century was provided by a shifting and often interlocking network of government organisations and policies, aristocratic patronage, and private enthusiasm and enterprise. The grand vision which Banks proposed for Kew became blurred after his death in 1820. In 1841 it was given to the nation, and placed under the control of a government department, the Commissioners of Woods and Forests, with William Hooker as director. This greatly strengthened its role, and gave it an official status rivalling the natural history departments of the British Museum. When Richard Owen was appointed their superintendent in 1856, he and Hooker would confront each other from separate dominions, and from divergent scientific standpoints. Henry de la Beche, secretary of the Geological Society, was commissioned in 1832 to add geological data to the Ordnance Survey map of Britain. He became director of the Survey in 1835, and proceeded to found – and secure official funding for – the Museum of Economic Geology in Jermyn Street. This was housed in a prestigious building with frontage on Piccadilly, and opened in advance of the Great Exhibition. His next achievement was to set up a Government School of Mines 'and of Science applied to the Arts'. Beche was succeeded as head of the Ordnance Survey by the expansive socialite and geologist Sir Roderick Murchison (all these great men of science were knighted), best known for his work on the Silurian system. By mid-century the slightly haphazard, informal arrangements had been formalised and institutionalised.

Parallel to the institutions were the learned societies. The Linnean, founded in 1788, was the oldest, and began to publish its *Transactions* in 1791. The Geological Society of London began in 1807; in the 1820s and 1830s its discussions were notoriously heated, as proponents and opponents of the catastrophist theory did battle. Lockhart, editor of the *Quarterly Review*, regularly attended meetings, explaining that, although he did not care for geology, 'I do like to see the fellows fight'. The Zoological Society of London was founded in 1826, the Entomological Society in 1833. Other intellectual centres also fostered societies, for example Edinburgh with its Botanical Society in 1836. The transactions and proceedings of many of these bodies were published, and there were, too, less specialised but equally important journals and magazines, such as John Claudius Loudon's *Magazine of Natural History* and *Gardener's Magazine*, all disseminating information and new discoveries, and providing platforms for intellectual reputations, or sources of additional income. Near the top of the pyramid sat the formidable national bodies, the British Association for the Advancement of Science and the Royal Society.

The universities provided a certain, but initially fragile, underpinning framework. Robert Jameson, appointed Professor of Natural History at Edinburgh in 1804, introduced field classes in mineralogy, based on his own experience of German teaching. He would lecture in the open air at Arthur's Seat, and his students included over the years Robert Grant, the first Professor of Zoology at London, Charles Daubeny, Professor of Chemistry and then of Botany at Oxford, and Charles Darwin (who thought Jameson's lectures incredibly dull). William Buckland, Reader in Mineralogy at Oxford, similarly led his students into the fields, formally dressed and in academic gown, or rode with them to the top of Shotover Hill, where he might produce problematic bones or fossils from the depths of a blue bag. Adam Sedgwick, the Cambridge mathematician, was in 1817 appointed Woodwardian Professor of Geology, a subject he cheerfully admitted he knew nothing about. However, he put this ignorance right – ('Hitherto I have never turned a stone: now I shall leave no stone unturned') – and led large parties of students on mounted tours across the Fens, ending with a lecture on drainage delivered from the roof of Ely Cathedral. Systematic study of the sciences at the universities took some time to gain recognition. In 1834, the Chair of Natural History at King's College, London was scrapped because of the lack of interest, and in 1842 Edward Forbes, the Professor of Botany, had to find a second job with the Geological Society to supplement his meagre pay. Huxley, the archetypal professional scientist, was in the 1850s forced to scramble a living together from an accumulation of posts and lectureships. But the university context had many facets: it was through the Cambridge network that Professor Henslow made the inspired recommendation of his former pupil, Darwin.

The journeys of many, perhaps most, of the scientific explorers were part of the imperial process. Darwin was a passenger on HMS *Beagle*, Joseph Hooker a 'scientific' assistant surgeon on the *Erebus* as Thomas Huxley was on the *Rattlesnake*. They catalogued the natural world and its history as inexorably as the naval chart-makers surveyed the oceans. These scientists could be seen as another species of map-maker, going out from Europe to track down every living thing, and bringing specimens back for the admiration and edification of the less adventurous members of the tribe.

During the course of the nineteenth century, however, a gradual but highly significant change took place. The Linnean system, product of the eighteenth century of rationality and order, supposed a fixed order of creation. Everything in existence had been created: Genesis chapter 1 ruled, more or less. If you interpreted the myth of creation freely, you could hold a position in which each day represented an epoch, and so advance from the increasingly untenable idea of a once-for-all creation to a stage-by-stage theory, in which God worked through a series of successive acts: the evidence, in the form of plants and animals, fossils and rock strata, lay all around. The bands of naturalists who fanned out across the world, Linnaeus's dedicated followers, were working to a pre-ordained plan, a giant jigsaw puzzle with hundreds of thousands of missing pieces, but one whose outline had been drawn on the cover of the box, by Linnaeus or Cuvier. Theirs was a Eurocentric world, conceived by Western, European minds, whose centres were Uppsala, Paris, Berlin, Vienna, London; and they duly brought their specimens back to the capitals of Europe for classification, where they piled up in the basements of museums, or stocked the zoos and botanical gardens. This was an attempt to catalogue everything within a universal filing system, so that they could be studied through encyclopaedic works like the forty-four volumes of Buffon's *Histoire naturelle*. The emphasis was on identification and classification, the what rather than the how, though there were disturbing features in the theories of Lamarck, which suggested that living creatures might somehow will themselves to adapt and change. But the tendency was far more towards contented inspection and observation. Gilbert White could record the wonders of his garden and the countryside at Selborne like an Adam blissfully contented with his ordered paradise, simply intent on getting to know it better. The alternative, opposing theory of evolution was like distant thunder, rumbling on the horizon, until Charles Darwin and Alfred Wallace independently achieved their breakthrough. The publication of *The Origin of Species* in 1859 signalled that the final and essential act of revision had taken place. The storm then burst, and the old world order could be accommodated no longer. Life was in flux, not fixed.

In a room in Darwin's house at Down, in Kent, hangs a large painting of Alfred Russel Wallace. In the background, there is a thatched hut on stilts, Wallace's temporary home on an island in the Malaysian Archipelago, Sulawesi, perhaps, or Gilolo; in the foreground, the lanky figure of the naturalist sits at a table, working on the body of a bird of paradise, while all around his head fly living specimens of the wonderful creatures he had travelled so far to find. This figure represents the scientist in the forest, investigating nature in the wild. Wallace seems in this image a kind of St Francis in reverse; though he was essentially gentle in manner and spirit, the Victorian naturalist's first duty was to shoot and trap and skin and stuff and pickle, fulfilling through capture and slaughter the role of an imperial Adam. The contrast with the very English context of Darwin's home, Down House, is startling. At Down, surrounded by his books and specimens, Darwin thought and wrote, and paced the Sandwalk, and carried out his investigations on earthworms and pigeons, sorting out the momentous implications of his theories in the midst of a superficially orthodox pattern of life, while Wallace, our man in Ternate, tracked down the bird of paradise in his solitary quest for Eden. The English setting is wonderfully understated; the village of Down itself is approached by twisting lanes just inside the perimeter of the M25; the house, badly in need of a coat of paint, looks slightly neglected, as though an unsuccessful nursing home or preparatory school had moved out with a few years of the lease still to run. Yet this is one of the intellectual powerhouses of Europe, as resonant in its implications as Rousseau's Arcadian shrine at Ermenonville, or Jung's lakeside tower at Bollingen on the Zürichsee.

The investigation of nature was the great nineteenth-century work. When Spinoza referred to the cosmos as 'God or Nature', he was signposting the way to forbidden territory. The whole thrust of seventeenth- and eighteenth-century philosophy, Descartes, Hume, Kant, pushed further down the track, to search for the understanding of the world within the world, rather than outside it; to investigate the physical world, and man as part of that world. For that great purpose, Europe could no longer be self-sufficient. Ever since the 'discovery' of the New World, in a sense, the European mind and imagination had been in training for the wilderness, going through a few preliminary exercises in its own back yard, in the Harz mountains, the Lake District, the Alps, the forest of Fontainebleau; but these were small-scale forays, merely enlivened by the odd wolf or viper. The world outside Europe was increasingly available; and going there, being there, becoming a part of another rhythm of life, of nature wholly untamed and uncivilised (and coming back to reap the reward) was at last a workable proposition. Joseph Banks deliberately chose a round-the-world trip for his definitive Grand Tour – 'any blockhead can go to Italy'. It was not long, of course, before it was possible to experience even the paradise of the South Seas vicariously. At Hawkstone in Shropshire, for example, where the whole gamut of the sublime was supplied to the reasonably energetic, including Hermitage, Menagerie, Grotto and Awful Precipice, a 'Scene at Otaheite' was added, derived from Captain Cook's tour.

There is here a broad analogy with the history of thought. Once traditional explanations of God have been rejected, in which authority comes from 'above' and in which a sharp distinction is drawn between the higher ideal and the natural, the thinker has to plunge into what is left 'below'. 'You have to go through inner turmoil; you have to descend into the primal chaos,' as Don Cupitt described the process in *The Sea of Faith*, 'and there await the remaking of meanings and the emergence of a new reality.' The scientists, literally, went downwards, charting the earth's history through the rock strata and the fossil deposits; and the naturalists plunged into the unknown and the far-flung. The ordered, man-made landscape of the picturesque, even one supplied with grottoes or chasms, became simply a preliminary experience, leading to the unfamiliar territory of the sublime: a real, not a simulated, exposure to the dark mysteries of the earth, the awful grandeur of jagged mountains, the distant sources of rivers, the chaos of ice-floes, and the exotic plants and primitive inhabitants of the tropical forest.

In social and cultural terms, the scientific travellers were representatives: a new form of the searcher for truth, a kind of pilgrim (as Conrad sardonically labelled the investors and entrepreneurs in *Heart of Darkness*). Like Gawain on his epic journey to confront the Green Knight, they battled through desert wastes, across torrents, over mountains, on their various quests. The image of the forest, tropical, primeval, seething with life, and its accompaniments of river and chasm, emerges as the key location, both a challenge and an end in itself. The tropical forest becomes the new Garden of Eden, an environment in which all forms of life present themselves for inspection. These forms include scarcely known, unsettling species, such as the orang-utan and the gorilla, and disturbing orders of wild men. The forest, from which Isabela Godin emerged, becomes not only an extreme experience to be endured, but the primal and true source of understanding, the site of the tree of knowledge. Poetically, this sense of searching for the creative force, and source, of the world is expressed in Coleridge's 'Kubla Khan', which reaches out from its physical English context of the Quantocks and the Cheddar Gorge towards a wilder, more elemental landscape:

> And from this chasm, with ceaseless turmoil seething,
> As if this earth in fast thick pants were breathing,
> A mighty fountain momently was forced:
> Amid whose swift half-intermitted burst
> Huge fragments vaulted like rebounding hail,
> Or chaffy grain beneath the thresher's flail;
> And 'mid these dancing rocks at once and ever
> It flung up momently the sacred river.

Coleridge, at the close of the eighteenth century, seems to anticipate the search for the creative mechanism which powered the earth.

Blake expressed the power and beauty of the natural world:

> The pride of the peacock is the glory of God.
> The lust of the goat is the bounty of God.
> The wrath of the lion is the wisdom of God.
> The nakedness of woman is the work of God.

This 'Proverb of Hell' celebrates the centrality of energy and sexuality as a principle of life, a poetic anticipation of the idea of natural selection. In the late eighteenth century, the Creation was a favoured subject for illustration in painting, with Milton's

Paradise Lost as a frequent choice of textual source. Blake's painting, *The Temptation of Eve*, refers to both Genesis and Milton as well as conveying his own personal theology, but its main motifs of massive overarching tree, waterfall and river, the moon in the night sky, form a framework for the towering coils of the serpent, and the shining body of Eve, her right arm raised not towards the apple but in celebration; in contrast, Adam lies asleep, labourer's spade by his side. The whole image suggests the act of creation, and of an ascent as much as any notion of the Fall. The ambivalence of the clothed and sophisticated Western world towards sexuality, and particularly towards sexuality within a primitive way of life, is a frequent preoccupation of travellers: Isabela Godin appearing naked from the forest has a counterpart in Mrs Thompson, the young Scottish woman who was nurtured for five years by the Australian aboriginals. Barbara Thompson had run off with a sailor at the age of fifteen, and been rescued from drowning by Torres Strait natives. She appeared to the crew of the *Rattlesnake*, 'a white woman disfigured by dirt and the effect of the sun on her almost uncovered body', and introduced herself with an echo of Genesis: 'I am a Christian – I am ashamed.' These women were like myths brought to life; and the imaginative re-picturings and re-readings of creation were echoed in the narratives of the scientific explorers as they moved among the forests of the night.

At the very beginning of the nineteenth century Alexander von Humboldt and Aimé Bonpland made an expedition to South America which served as a model for their naturalist successors. Humboldt, described as 'the last great universal man', felt strongly that he was a direct inheritor of the Cook tradition, for he had benefited from the teaching and experience of Georg Förster, the naturalist who had replaced Banks on Cook's second great voyage; he had, of course, read La Condamine's accounts of America, while Bonpland had met a survivor of that expedition in Paris. Humboldt enjoyed inherited wealth, and could finance himself; the two men spent five years on their travels. Humboldt began by exploring the upper waters of the Orinoco and the Casiquiari Canal; later, by contrast, he climbed higher in the Andes on the volcanic Chimborazo than anyone before him. When he returned to Europe, he produced volume after volume, and his *Personal Narrative of Travels to the Equinoctial Regions of the New Continent during the Years 1799–1804* became a naturalist's bible, with its potent evocation of the luxuriant forest. The images of Humboldt and Bonpland in Eduard Ender's painting convey vividly the idea of the new scientists in their domain: a rude shelter, a rough-hewn table, scientific instruments in profusion, paper to record and

1. Alexander von Humboldt, and Aimé Bonpland contemplating the riches of the Orinoco (Engraving, O. Roth).

preserve the botanical specimens, and a cornucopia of plants spilling off the table and on to the ground; the light shining on the two young men, representatives of France and Germany, of the Enlightenment – new Adams, monarchs of all they survey. At first sight, they seem to have the territory to themselves, though the number of chests and boxes indicates the need for porters, and in the right foreground is a carrying frame; only at second glance can one pick out a diminutive figure in the background, an Indian at the drudge's task of physically collecting specimens. Roth's engraving confirms the image. Humboldt and Bonpland catalogue the fruits of the forest, surrounded by specimens and corpses, while their Indian helpers wait outside, part of the forest which is being stripped for inspection.

Humboldt and Bonpland doubled the number of plants known to the Western world. They were the trailblazers through the world's forests. Humboldt's descriptions were constantly in Darwin's thoughts. The great triumvirate of Amazon naturalists, Bates, Wallace and Spruce, followed in Humboldt's footsteps – literally, in the case of Wallace and Spruce – as they quartered the forest of the Amazon basin and the Andean foothills.

The idea of the scientific traveller, particularly the figure of the naturalist in the forest, lies at the centre of this book, together with the way in which their conclusions were translated, and sometimes distorted, into other cultural forms; since the period is the nineteenth century, the chief form is the novel.

Sometimes the distance between a scientific account and fiction is negligible; a bookseller's list of 1817 advertises, between *The Fourth and last Canto of Childe Harold* and *Rob Roy*, the following:

> *On the Possibility of approaching the North Pole*, by the Hon. D. Barrington;
> *Voyage of His Majesty's Ship Alceste along the Coast of Corea to the Island of Lewchew, with an account of her subsequent Shipwreck*, by John McLeod, Surgeon of the *Alceste*;
> *Narrative of an Expedition to explore the River Zaire, usually called the Congo, in South Africa, in 1816*, under the direction of Captain J.K. Tuckey, R.N., &c. &c.

These particular establishment productions have a reassuringly official and objective ring about them; but many of the later travels and journals have a far more reader-conscious style and construction, and the line between 'fact' and 'fiction' is often blurred, as in, for example, Paul du Chaillu's accounts of the gorilla in *Adventures in Equatorial Africa*.

I received my first knowledge of the rest of the world from fiction: from the covers of Henty's novels, from Kipling and Conrad. The antidote to my imperial education was a year in West Africa, passed almost entirely on the edge of the Niger delta. I was ill-prepared, mentally and physically: a visit to a tropical outfitters in Soho left me clutching a pair of goatskin mosquito boots and some baggy khaki shorts; I narrowly escaped being sold a portable bath 'for going to bush'. (The boots were invaluable, practical and comfortable, but did nothing for my image.) I can still remember the smell of Africa, sweet and heavy in the humid morning at Lagos airport; and the long drive by battered Volkswagen to the place where the tarred road came to an end, and turned into a rutted red scar; and my first view of the school compound, surrounded by bush and swamp and forest. At half-term, I bicycled for twenty miles to the home village of some of the boys in my class, or rather to a village where we left the bicycles before setting off in a canoe down some tributary of the Niger. I could not conceive of being further from England, or that life could be more straightforward and immediately enjoyable, surrounded by apparently solid green swathes of vegetation, with either a fierce sun or a rain heavier than I had ever experienced beating down. Everything seemed possible.

On a brief visit to Ibadan, I met a young Nigerian writer at the university. He asked me what I was doing: 'Teaching English.' 'Oh,' he replied, 'something tribal.' Chinua Achebe's great novels *Things Fall Apart* and *No Longer at Ease* completed the adjustment, or at least made me conscious of the need for one. The imperial glamour faded; the power of the tropical forest grew, context for a sense of life overwhelmingly fertile, where plants unfolded almost before your eyes and where the insects and snakes and lizards had a disconcerting habit of ignoring any man-made boundaries between indoors and outside. To live in the forest climate for a year at least allows one to grasp some sense of the scale of achievement of the scientific travellers, and especially the later independent, self-financed and often solitary naturalists, a Spruce or a Wallace, in their persistent search for knowledge. The selection, although I hope it is rational, is also personal, reflecting my own tamer journeys, which have crossed the tracks of some of these people: for a winter, I lived in the same North Yorkshire village to which Spruce retired; in the woods between my home in Swaffham Bulbeck and Bottisham, Darwin hunted for beetles in the company of Leonard Jenyns.

This book revisits some of the expeditions which the Victorian scientific travellers undertook, and considers how their journeys and writing helped to reshape a view of the world. One common factor they shared, obvious but crucial, is the way their travels were extended in time: they spent, on the whole, years rather than months away from home, from 'civilisation'. This holds true of the government-sponsored scientists, like Huxley, locked into the routine of a survey ship, sweating in his cabin as he read Dante's *Inferno*. It is even more significant with the land-based naturalists, such as Wallace, Bates and Spruce, who lived for years in a rhythm and landscape quite alien to their native land's, adapting their habits to fit in with their surroundings, and turning their intelligence not just to the plants and insects and birds around them, but to the human races who formed part of the spectrum of life. They were, almost incidentally, the first ethnographers, recording the languages, customs and beliefs of the indigenous peoples among whom they moved and on whom they relied.

In 1853 in a London theatre, two years after the Great Exhibition which proclaimed the achievements of modern industrial Britain, a group of Zulus was put on show to the public, a live demonstration of primitive behaviour to which Dickens reacted in disgust, not so much at the spectacle, or the fact of the spectacle, but at the concept of the ignoble savage which he saw at its heart. From their extended observation in the Amazon and the Malaysian Archipelago, Spruce and Wallace would suggest a very different and infinitely more complex idea of the civilised and the savage, one which questioned and challenged the comfortable duality of the Western world. Nature, and the forest, tropical rather than temperate, and everything that grew and existed in it, were placed in the foreground as never before.

1 *The Scientists of the Survey*

THE SYSTEMATIC CATALOGUING OF the world was a monumental undertaking. It was an attempt to classify every existing plant, animal and insect; and, largely by means of the geologist's hammer, to uncover evidence which would establish the age of the earth, and help to define the major phases of its story. Ironically, this large-scale quest contained a terrifying shock. Once sufficient data had been assembled, it would provide evidence for a new world view, one in which Europe could not be sustained as the centre of civilisation, in which to be white and Christian could no longer be seen as the highest work of God, in which even the existence and idea of God was for many people called into question. The young Charles Darwin, packing reference books and hammer and collecting gear on HMS *Beagle* in the summer of 1836, was as yet wholly innocent of his role. The *Beagle* was bound on a survey voyage to make the seas safer for British merchant ships, and had three Westernised converts on board to deliver back to Tierra del Fuego. The presence of Darwin was a kind of inspired afterthought: 'Capt. F[itzRoy] wants a man (I understand) more as a companion than a mere collector & would not take any one however good a Naturalist who was not recommended to him likewise as a gentleman'. Darwin's correspondent was Professor Henslow, his Cambridge Professor of Botany, who even contemplated making the journey for himself, before pushing the opportunity in the way of his brother-in-law, the Reverend Leonard Jenyns, who declined it. Darwin's father needed a good deal of persuading; uncle Josiah Wedgwood pointed out that 'the pursuit of Natural History, though certainly not professional, is very suitable to a Clergyman'. So, effectively, Darwin was diverted from his planned destination in a country rectory, and launched towards an unsuspecting world. In his tracks there followed a band of successors, men such as Bates, Wallace and Spruce, whose cumulative work would endorse and amplify the results of his journey. They were the last of the old travellers, or the first of the new.

Darwin, discussing the geology lectures of Adam Sedgwick

which he had attended at Cambridge, commented: 'It strikes *me* that all our knowledge about the stratum of our Earth is very much like what an old hen wd know of the hundred-acre field in a corner of which she is scratching.' The Victorian scientists set out to explore the whole hundred-acre field, leaving their own over-populated and too familiar corner often for years at a stretch, and bringing back, cargo by cargo, the riches and knowledge which they found on their travels. Finally, they used it to explain the structure and development of the earth and everything on it.

They travelled out, for the most part, as orthodox Christians, representatives of one of the most rigid and regimented societies in Europe: white European males, class conscious, racist, chauvinistic. Yet because they were, almost by definition of their business, slightly unorthodox as individuals, they were subject to change. The process itself changed them. They did not go out primarily as naval officers, as traders, as conquerors or prospective settlers, or even as missionaries, though some part of those roles often attached itself to their expeditions and dealings. They went, principally, as enquirers and observers. Their cumulative experience would alter the West's image of the rest of the world, and begin to shift the inherited understanding of its own society and beliefs.

Inevitably, Darwin and his successors saw the tropical forest, and the other world, partly through the eyes of previous European explorers and artists, and partly in terms of the contrast with their own experience of nature. Darwin had already been steeping himself in Humboldt, in preparation for a projected expedition to Tenerife. He wrote to his friend and cousin William Fox from Rio de Janeiro in May 1832: 'But when on shore, & wandering in the sublime forests, surrounded by views more gorgeous than even Claude ever imagined, I enjoy a delight which none but those who have experienced it can understand – If it is to be done, it must be by studying Humboldt.' Darwin took Humboldt's *Personal Narrative of Travels to the Equinoctial Regions of the New Continent during the Years 1799–1804* with him on his voyage, a present from Henslow. Humboldt's narrative supplied the lens through which the wonders of the equatorial forest could be perceived. At its best, botanising in the tropics was a walk through the garden of Eden, even if it was a rather long one: Darwin was only five months into a journey which had another four and a half years to run. 'I suppose,' he confided to Fox, 'I shall remain through the whole voyage, but it is a sorrowful long fraction of ones life; especially as the greatest part of the pleasure is in anticipation. I must however except that resulting from Nature-History; think when you are picking insects off a hawthorn hedge on a fine May day (wretchedly cold I have no doubt) think of me collecting amongst pineapples & orange trees; whilst staining your fingers with dirty blackberries, think & be envious of ripe oranges.' Fox, writing back from Epperstone Rectory in the tame Nottinghamshire countryside, pondered on the contrast between himself and Darwin surrounded by the wonders of creation: 'I pottering in a Hedge Rows to watch the proceedings of a Whitethroat & you surrounded by the Noble Trees of a S. American Forest with every luxury of vegitation & life around you.'

Darwin never lived in the countries he visited for any appreciable length of time; but the idea of the tropical forest was central to his imagination. He was full of anticipation about his first proper expedition up-country – with a merchant, travelling from Rio de Janeiro to visit his large estate at Rio Macao: 'I shall thus see,' he wrote to his sister Caroline, 'what has been so long my ambition, virgin forest uncut by man & tenanted by wild beasts.' Then – knowing what that description, even with the 'tenanted' metaphor, might do to his susceptible sisters – he added, half-jokingly: 'You will all be terrified at the thought of my combating with Alligators & Jaguars in the wilds of the Brazils.' So, having raised alarm, he immediately cancelled any suggestion of danger: 'The expedition is really quite a safe one, else I will wager my life, my host & companion, would not venture on it.' Caroline should get hold of a copy of a French engraving, *La forêt du Brésil* – 'it is most true and clever.' He wrote to Henslow about this same trip: 'Here I first saw a Tropical forest in all its sublime grandeur. – Nothing, but the reality can give any idea, how wonderful, how magnificent the scene is. . . . I never experienced such intense delight. – I formerly admired Humboldt, I now almost adore him; he alone gives any notion, of the feelings which are raised in the mind on first entering the Tropics.'

Darwin at this point was very conscious of his inexperience; a great source of perplexity, he confessed to Henslow, was whether he noted 'the right facts & whether they are of sufficient importance to interest others'. But in collecting he could not go wrong, and his hammer was his trusted companion. His short trip to Wales with Sedgwick in the summer of 1831, a rapid induction course, had given him confidence: they had even found some mammal bones in caves near St Asaph. He promised Henslow that geology and the invertebrate animals would be his 'chief object of pursuit' throughout the voyage. But everything interested him; and he was exhilarated to find that small insects had so far been ignored by the collectors: 'I tell Entomologists to look out & have their pens ready for describing'; he was 'red-hot' with spiders. Again, Darwin

2. *La forêt du Brésil*, the engraving Darwin thought 'most true and clever'. (Engraving, Moritz Rugendas; Bodleian Library, Oxford)

returned to the impact which the tropical rain forest had made on him. 'I well know the glories of a Brazilian forest,' he wrote to his sister Catherine. 'Commonly I ride some few miles, put my horse & start by some track into the impenetrable – mass of vegetation.' But the instincts and habits of the English gentleman died hard; Darwin was only a temporary visitor. 'Whilst seated on a tree, & eating my luncheon in the sublime solitude of the forest, the pleasure I experience is unspeakable.' Luncheon over, it was back to business: 'The number of undescribed animals I have taken is very great – & some to Naturalists, I am sure, very interesting. – I attempt class after class of animals, so that before very long I shall have notion of all.' That sense of all-embracing completeness which is characteristic both of the age and the man is hinted at in the modest phrase, 'I shall have notion of all'. He added, ' – so that if I gain no other end I shall never want an object of employment & amusement for the rest of my life'. The work ethic, nurtured in Unitarian, nonconformist England, drove Darwin forward, and his self-prophecy proved triumphantly correct. The collection and the ideas he derived from his long field trip into the natural South

American laboratory provided him with material for the rest of his life.

Writing in his *Journal of Researches* about his voyage in the *Beagle* more systematically than in his private letters (the book eventually appeared in May 1839), Darwin expanded upon the delight of the naturalist who, for the first time, was able to wander by himself in a Brazilian forest.

> The elegance of the grasses, the novelty of the parasitical plants, the beauty of the flowers, the glossy green of the foliage, but above all the general luxuriance of the vegetation, filled me with admiration. A most paradoxical mixture of sound and silence pervades the shady parts of the wood. The noise from the insects is so loud, that it may be heard even in a vessel anchored several hundred yards from the shore; yet within the recesses of the forest a universal silence appears to reign. To a person fond of natural history, such a day as this brings with it a deeper pleasure than he can ever hope to experience again.

For Darwin, as for successive visitors to the Amazon, the luxuriance and fecundity of the rain forest marked the existence of a new world. England was tame, and relatively known. You could claim exhaustion after wandering through the Welsh mountains to Dolgellau or plunging about in Whittlesea Mere in search of beetles, as he had done as an undergraduate; but there was limited kudos. There were too many botanising parsons, and parsons' wives and parsons' daughters. That was apprentice work, which he had taken seriously enough. He had tossed about on the Firth of Forth, to scrutinise what the Scottish fishermen trawled up from the sea-bed, and employed people in Cambridge to rummage about in the barges on the Cam. But in England you could only do, more thoroughly, what other naturalists were doing or had done, building on the meticulous work of their European predecessors, Ray, Linnaeus, Buffon, adding inexorably to the record of nature's variety. By extending the rock-smashing and beetle-hunting to the rest of the world, Darwin was making the leap forward towards tackling the great questions: how and why.

Darwin omitted writing very much about his first American landfall, Bahia or San Salvador, near the eastern bulge of the continent, on his voyage out, rightly confident that his meticulous captain, FitzRoy, would complete the survey by returning there. By the time they called at Brazil a second time, in August 1836, he was searching for words with a growing sense of impatience and homesickness, conscious too, perhaps, that he was viewing the tropical forest for the last time:

> Who, from seeing choice plants in a hot-house, can magnify some into the dimensions of forest trees, and crowd others into an entangled jungle? Who, when examining in the cabinet of the entomologist the gay exotic butterflies and singular cicadas, will associate with these lifeless objects, the ceaseless harsh music of the latter, and the lazy flight of the former, – the sure accompaniments of the still, glowing noonday of the tropics? It is when the sun has attained its greatest height that such scenes should be viewed: then the dense splendid foliage of the mango hides the ground with its darkest shade, whilst the upper branches are rendered from the profusion of light of the most brilliant green.

Darwin expressed both the essential difference of the tropics, in terms of light and colour, and the feelings and sensations which these produced; and yet language failed him, other than the language and constructs of Europe. 'I have said that the plants in a hot-house fail to communicate a just idea of the vegetation, yet I must recur to it. The land is one great wild, untidy, luxuriant hot-house, made by Nature herself, but taken possession of by man, who has studded it with gay houses and formal gardens.' And by man, Darwin seems to imply European man, moving inexorably south and west and east in successive waves of colonisation, taming and organising the tropical world into an exotic centrally heated extension of Western civilisation.

Reflecting on the totality of his four-year voyage in the last chapter of his *Journal*, and taking the opportunity of a new 1845 edition to make some revisions, Darwin voiced two strong but contradictory judgements. The society he had observed in South America was repugnant to him, because it was based on slavery. 'I thank God I shall never again visit a slave country,' he wrote.

> To this day, if I hear a distant scream, it recalls with painful vividness my feelings when, passing a house near Pernambuco, I heard the most pitiable moans, and could not but suspect that some poor slave was being tortured, yet knew that I was as powerless as a child even to remonstrate. . . . I have seen a little boy, six or seven years old, struck thrice with a horse-whip (before I could interfere) on his naked head, for having handed me a glass of water not quite clean; I saw his father tremble at a mere glance from his master's eye. These latter cruelties were witnessed by me in a Spanish colony, in which it has always been said that slaves are better treated than by the Portuguese, English or other European nations.

But the same man whose moral sense was outraged by the inhumanity of slavery, the product of civilised Europe, was appalled by his first sight of natural man, of man in an apparently savage state. The episode came as a profound shock. Darwin, for all his honorary, 'supernumerary' status, was very much part of the establishment; on board a naval ship, subject to naval orders, and enjoying all the network of contacts and privileges which followed the Beagle and her captain on what was an imperial mission: dinners with governors, official receptions. Among his last engagements at Montevideo had been a grand ball to celebrate the re-establishment of the President, and a performance of Rossini's La Cenerentola. This was a far cry from the rugged desolation of Tierra del Fuego. Besides, he knew some Tierra del Fuegans already, for on board were three natives, Jemmy Button, York Minster, and a young girl, Fuegia Basket: they had originally been taken as hostages, and FitzRoy was anxious to return them, as Christians, with an English catechist from the Church Missionary Society to help establish a Christian settlement. The three had lived in Walthamstow, and came laden with well-intentioned but misconceived presents from the parishioners there. The young Jemmy Button was a particular favourite of Darwin's, and his long conversations with him stayed in his mind.

The first sight of his Yahgan kinsmen, however, presented Darwin with a very different image; as the ship entered the Bay of Good Success, a group of Yahgans, partly concealed by the entangled forest on a wild point overhanging the sea, sprang up, waved their tattered guanaco-skin cloaks – the only garments they had over their dirty coppery-red skin – and, their long hair blowing about, 'sent forth a loud and sonorous shout'. As he described it to his friend Fox, there was something inconceivably wild about their countenances; standing on a rock, the naked Fuegian 'uttered tones & made gesticulations than which, the crys of domestic animals are far more intelligible'. The next morning, Darwin went on shore with a party to establish communications: 'It was without exception the most curious and interesting spectacle I had ever beheld. I could not have believed how wide was the difference, between savage and civilised man. It is greater than between a wild and domesticated animal, in as much as in man there is a greater power of improvement.' And these Fuegians were a superior race to the wretches to be found further west. They reminded Darwin of the devils in Weber's *Der Freischütz*. At another anchorage, Darwin saw a group he described as 'the most abject and miserable creatures' he had anywhere beheld: they were, even full-grown women, quite naked; one woman, suckling a recently born child,

3. The Beagle Channel, Woolya Cove. (R. FitzRoy, *Narrative*, 1839)

stayed in her canoe whilst sleet fell and thawed on her bosom and on the skin of her baby. 'These poor wretches were stunted in their growth, their hideous faces bedaubed with white paint, their skins filthy and greasy, their hair entangled, their voices discordant, their gestures violent and without dignity. Viewing such men, one can hardly make oneself believe they are fellow-creatures, and inhabitants of the same world. It is a common subject of conjecture what pleasure in life some of the less gifted animals can enjoy: how much more reasonably the same question may be asked with respect to these barbarians.' They slept naked on the wet ground coiled up like animals; their life was spent either picking shellfish from the rocks at low tide, or diving for sea-eggs or sitting for hours in a canoe fishing – a diet supplemented by a few berries and fungi. In times of famine, he wrote authoritatively but almost certainly incorrectly, there was 'cannibalism accompanied by parricide'. (In the second edition of the *Journal*, Darwin expanded on these imagined horrors, citing Jemmy Button and the reports of a sealing-master, Low, as evidence that in times of hunger the Fuegians killed and devoured their old women before they killed their dogs, on the grounds that 'Doggies catch otters, old women no'. Low's informant even suggested the victims were smoked and roasted alive. This may have been the sort of spine-chilling tale the Fuegians thought their audience would like to hear; but Darwin was quite prepared to believe the worst.)

When the *Beagle* left Tierra del Fuego, the last native they met was Jemmy: they had left him plump, fat, clean, well dressed – a man who had been presented to King William IV at the English Court; now he was a thin haggard savage, with long disordered hair, naked except for a bit of blanket; but he told them he had enough to eat and was not cold, and had no wish to return to England; the main reason, Darwin guessed, was that he now had a young, nice-looking wife. Darwin convinced himself that Jemmy Button would be as happy as, perhaps even happier than, if he had never left his own country. But he remained perplexed about the Fuegians. They were too democratic, living in a state of 'perfect equality'; even a bit of cloth given to one would be torn in shreds and distributed, so no one could become richer than another; and until some sort of property could be established, how could a chief arise and demonstrate his superiority and increase his power? In the wake of the *Beagle*, presumably, would follow commerce, property, and all the other consequences which might raise the Fuegians from their primitive state. (Mathews, the English catechist, had already been rescued from what was, at this point, a clearly hopeless endeavour.)

Yet there was a very long way to go. At Montevideo in 1832, Darwin had received the second volume of Lyell's *Principles*, which argued against transmutation of species: to Lyell, the idea of any connection between man and ape was anathema. It was easier, perhaps, to conceive of man emerging from less highly developed species than to contemplate a race which embraced the Victorian naval captain and philosopher on board the *Beagle*, and a wild-haired naked Fuegian crying like an animal from a rock.

So strong an impact did this whole encounter make that Darwin returned to it in his conclusions to the *Journal*:

> Of individual objects, perhaps nothing is more certain to create astonishment than the first sight in his native haunt of a barbarian, – of man in his lowest and most savage state. One's mind hurries back over past centuries, and then asks, Could our progenitors have been men like these? – men, whose very signs and expressions are less intelligible to us than those of the domesticated animals; men, who do not possess the instinct of those animals, nor yet appear to boast of human reason, or at least of arts consequent on that reason. I do not believe it is possible to

paint or describe the difference between savage and civilised man. It is the difference between a wild and tame animal; and part of the interest in beholding a savage is the same which would lead every one to desire to see the lion in his desert, the tiger tearing his prey in the jungle, or the rhinoceros wandering over the wild plains of Africa.

That was the key question forming in Darwin's mind, with all the implications which lay behind it: 'Could our progenitors have been men like these?' In this passage Darwin seems to wish to suggest that they could not. He was still heavily under the influence of his bleak impressions of the native people of Tierra del Fuego. Naked, abject, miserable, they were the ignoble savages. Yet there was no reason to believe that the Fuegians were decreasing in number, so they must have a sufficient share of happiness to sustain their way of life. They were, for him, little better than animals: in fact, when he saw his first ape, Jenny the orang-utan, in London Zoo in 1838, the comparison was all in the ape's favour: 'Let man visit Ourang-outang in domestication, hear expressive whine, see its intelligence when spoken [to]; as if it understands every word said – see its affection. – to those it knew. – see its passion & rage, sulkiness, & very actions of despair; let him look at savage, roasting his parent, naked, artless, not improving yet improvable & let him dare to boast of his proud preeminence.'

Darwin's theory of the descent of man was already implicit. But if man should not boast of his pre-eminence, Darwin's belief in progress, founded on Christianity and the philanthropic spirit of the British nation, was still apparently unshaken. His experience in Tahiti had convinced him that the Christian missionaries had had a bad press, for he had seen strong evidence for their positive impact on the natives. The South Seas, and Australia, were rising stars. It was impossible for an Englishman to 'behold these distant colonies' without pride and satisfaction: 'To hoist the British flag, seems to draw with it, as a certain consequence, wealth, prosperity, and civilisation.' This was the voice of the establishment speaking in full-throated articulation, drowning out the disturbing cries of the naked savage.

In October 1836 Darwin returned home to Shrewsbury; he had been away for almost exactly five years, far longer than he had anticipated; but, as he stated firmly in the conclusion to the *Journal*, nothing could be more improving to a young naturalist than a journey in distant countries – by land, if possible, but if not, by sea. Darwin himself had spent more time on land than on the *Beagle*, but had not had the opportunity of extended detailed observation.

However, he argued, since a number of isolated facts soon becomes uninteresting, 'the habit of comparison leads to generalization', and the traveller has a tendency 'to fill up the wide gaps of knowledge, by inaccurate and superficial hypotheses'. Early in 1838, Darwin moved to lodgings in London, and in July, after sifting in his mind some of the evidence provided by the fossils of South America and the fauna of the Galapagos Archipelago he had visited in 1835, he began the first notebook on 'transmutation of species'. The subheading, *Zoonomia*, a glance in his grandfather Erasmus's direction, announced his commitment to evolution. For the next twenty or so years, he intermittently constructed a complete and profound hypothesis, but he did not declare it to the world until he was pressured to do so by the arrival of Alfred Wallace's paper in June 1858, sent to him from Ternate for an opinion.

The reasons for the delay invite speculation. At this distance, they seem wholly understandable. Darwin went out on the *Beagle* as an enthusiastic amateur, an inexperienced third choice behind Henslow and Jenyns. For five years he had observed, noted and conjectured, and had accumulated a vast store of facts and data; but he needed time, time for thought and reflection and continued research, to make sense of what he had seen. He needed, too, to plug the gaps, even gaps where his own collecting had been concerned – FitzRoy's collection of bird skins was used to fill out his own incomplete survey of the Galapagos finches and mockingbirds. Thanks to Henslow, who had published some of Darwin's letters at the meetings of learned societies, Darwin was already welcomed by the scientific establishment. In 1837 he read a paper to the Geological Society, and in January 1839 he was elected a Fellow of the Royal Society, a few days before his marriage to Emma Wedgwood. From this period, too, he began to suffer from occasional, and later persistent and acute, bouts of ill-health, which cut back his capacity for sustained work, extraordinary though his output still seems. There was the work on the *Journal* to be done, and on the collections; there was other specialised scientific writing, including his monumental research on barnacles. There were domestic responsibilities, though the Darwin and Wedgwood wealth ensured that he did not need to work for a living. All these factors stood in the way of his progress with the grand design.

But there were other strong reasons for discretion, even secrecy. At the same time as Darwin's clarity about the link between natural selection and evolution grew, he seems to have become equally clear about the shock that publication would cause. He wanted to protect his wife, an orthodox Christian; he had no wish to upset his friends, many of whom were Anglican clergy; he knew that the implications were profound – for religion, for science, for society. Even when *The Origin of Species* was published, he took care to sideline the position of man. In 1842 he wrote a sketch including these words: 'From death, famine, rapine, and the concealed war of nature we can see that the highest good, which we can conceive, the creation of the higher animals has directly come.' That was not something he could show to Lyell; even to Hooker he proceeded by cautious qualifications: 'I am almost convinced (quite contrary to the opinion I started with) that species are not (it is like confessing a murder) immutable.' During 1844 he expanded the sketch into an essay. He considered publication. The outcry which followed the anonymous, evolutionary *Vestiges of the Natural History of Creation* in 1844 was not encouraging: Robert Chambers, whom Darwin deduced was the author, claimed to have eleven reasons for remaining anonymous – his eleven children. In the end Darwin waited, though he left instructions to his wife to ensure publication in case of his death. He could share his ideas with carefully selected, trusted friends, such as Joseph Hooker and, much later, Asa Gray at Harvard; he could hint and test the water with the orthodox – though Leonard Jenyns did not take up his offer of a copy of the essay. In the 1850s he embarked on a large-scale, systematic *magnum opus*. Meanwhile, he went on piling up evidence about distribution and varieties in the peace of his country retreat at Down.

Two scientists closely associated with Darwin followed him in making significant voyages as young men, though in rather different circumstances. Where Darwin travelled as a self-financed companion, Hooker and Huxley held official naval posts as assistant surgeons. Joseph Hooker met Darwin briefly in 1839, and had read the proofs of the *Journal* – Lyell had shown his copy to Hooker's father, and Joseph slept with the proofs under his pillow. Just as Darwin had taken Humboldt with him, so Hooker had his presentation copy of the *Journal* on board HMS *Erebus* when he sailed with Ross to the Antarctic. During the four years of the voyage he collected in the Falklands and Tierra del Fuego – 'the great botanical centre of the Antarctic Ocean'. He found a welcome from Darwin waiting for him on his return in 1843, and, while he wrote up his findings in *The Botany of the Antarctic Voyage*, was drawn slowly into Darwin's confidence.

Hooker was always destined to be a botanist: his father was Professor of Botany at Glasgow and the owner of a renowned private herbarium, before his later appointment to Kew; his mother's father, Dawson Turner, was a distinguished amateur botanist. As his first biographer, Leonard Huxley, wrote, 'He did not so much learn botany as grow up in it.' As a boy Joseph Hooker would sit on his grandfather's knee, and gaze at the pictures in Cook's *Voyages*: 'The one that took my fancy most was the plate of Christmas Harbour, Kerguelen Land, with the arched rock standing out to sea, and the sailors killing penguins.' His father showed him a scrap of the brilliant green moss which had caught Mungo Park's eye when he was dying with hunger and exhaustion, and inspired him to keep going; and the young Hooker dreamed of crossing Africa to Timbuktu. By the beginning of 1839, Hooker had taken his university courses at Glasgow, had prepared himself, under his father's guidance, as a botanist and naturalist, and was on the verge of completing his qualifications in medicine. Captain James Ross was getting ready for his Antarctic expedition in search of the South Magnetic Pole, and Joseph Hooker was adroitly manoeuvred into the post of assistant surgeon and botanist on the *Erebus*.

Hooker was not entirely happy about his official position: he desperately wanted to be the official naturalist, so that he could have the indisputable first right to go ashore, and have more control over the collections and their publication. Ross appointed the surgeon, Robert McCormick, as zoologist. Ross told Hooker that a naturalist must be perfectly well acquainted with every branch of natural history, and 'must be well known in the world beforehand, *such a person as Mr Darwin*; here I interrupted him with "what was Mr D. before he went out? he, I daresay, knew his subject better than I now do, but did the world know him? the voyage with FitzRoy was the making of him (as I had hoped this exped. would me)."' Hooker petitioned everyone he could contact to change his status. He was, perhaps, apprehensive about competition from the more experienced McCormick, who had left the *Beagle* at Rio, riled by FitzRoy's preferential treatment of Darwin. In the end, he had to be satisfied with Ross's regarding him as the botanist to the expedition; and McCormick, who was interested almost exclusively in geology, proved a sympathetic and tolerant colleague. Together, they even encountered Darwin in the Charing Cross Road, and Hooker noted Darwin's 'animated expression, heavy beetle brow', and mellow voice. William Hooker visited the ship at Chatham, where it was fitting out; he wished that he had witnessed the conversation 'taking a more scientific and soberer turn', and that his messmates had paid more respect to the Sabbath. But he gave his son a beautiful chronometer watch, instruments and books, while Lyell added a copy of the first edition of *The Voyage of the Beagle*. Even though he was the official botanist he was sparsely equipped – 'not a single instrument or book supplied to me as a naturalist'. The government outfit consisted merely of twenty-five reams of drying paper for his plants, two botanising vascula, and two of 'Mr Ward's invaluable cases for bringing home plants alive'; for collecting, Hooker was forced to fall back on empty pickle jars, with rum from the ship's stores as a preservative.

The *Erebus* and the *Terror* sailed on 30 September 1839, on a voyage of almost four years which included three separate expeditions into the Antarctic Circle. There was not a great deal of material for a botanist in the Antarctic itself; but Hooker took every opportunity to collect in all the islands they visited, including Tasmania, New Zealand, Kerguelen Land, the Falklands, Hermite Island off Tierra del Fuego, the South Sandwich Islands; and he kept himself fully occupied at sea, when he was not working up the fruit of an island visit, by dissecting and drawing the marine creatures dragged up by the tow-net.

On the voyage out, the expedition stopped at Kerguelen – or Desolation – Island, sacred to Hooker because he had so often imagined it through Cook's eyes.

> From a distance the Island looks like terraces of black rocks; on which the snow lies, causing it to look striped in horizontal bands. On the melting of the snow, the flats appear covered with green grass and the hills with brown and yellow tufts of vegetation. The shores are almost everywhere bounded by high, steep precipices, some of frightful height, above which the land rises in ledges to the tops of the hills. The varied colour in the vegetation gave me hopes that the country might be rich in mosses, &c.

During their stay, Hooker doubled the number of known flora, and was delighted at finding the plants in a good state of flower and fruit. He went on several boating expeditions, but was dismasted on one and nearly swamped, which led Ross to ban such trips; so he rambled, generally on his own, in every direction from the harbour, several times starting before light, as the days were so short. It proved hard work to prise out the lichens; he had to hammer out the tufts, or even sit on them till they thawed. Seaweeds and lichens dominated – the latter appeared to 'form a greater comparative portion of the vegetable world' than anywhere else on earth; the rocks seemed to be painted with them, 'their fronds adhering so closely to the stone that they are with difficulty detached', and at the tops of the hills they assumed the appearance of a miniature forest on the flat rocks: the colours were wonderful, lilac, bright yellow, light red. The birds had no fear of man.

There was a beautiful 'Sheathbill': 'On one occasion I thoughtfully sat down on a stone and commenced whistling a tune when, on turning my head, I found I had unwittingly been performing an Orpheus's part, for upwards of twenty of these beautiful birds had gathered about me, and were gradually approaching, declining their heads and narrowly watching my motions, and would even perch on my foot, rocking their heads on one side.' The penguins, too, were so tame that they allowed Hooker to take them by the beaks.

Tasmania was the next stop, where Hooker met the Governor, Sir John Franklin, already a distinguished explorer, who would end his life in the search for the North-West Passage. Hooker's botanising here resulted, eventually, in his *Flora of Tasmania.* The first voyage towards the Magnetic Pole set off in November 1840; Ross reached within 150 miles of it, but progress was blocked by the ice barrier. Ross duly appropriated 'Victoria Land': on 12 January 1841, in the words of Cornelius Sullivan, the blacksmith of the *Erebus*, 'Captn. Ross went on Shore he took possession of the Land without opposition In the name of Queen Victoria – hoisted the British Colours Gave the Boats Crew an allowance of Grog with three hearty cheers for Old England'. Then it was back to Tasmania, and from there, via Sydney, to New Zealand, and the second attempt. In January 1842 the ships were battered in a terrible storm before they reached the Great Barrier Reef six miles further south than the previous year. The onset of winter forced them north and east, and while sailing towards the Falklands, in yet another fierce storm, the ships collided in attempting to avoid an iceberg. For three-quarters of an hour the *Erebus* rolled against the berg, before she could be cut free. They reached the safety of the Falklands on 6 April, and were based there, and at Hermite Island, to the west of Cape Horn, until December.

Hooker's first impressions of the Falklands were unenthusiastic. Kerguelen Land was a paradise in comparison to it, but he slowly became more interested in its botany, especially in the mosses and the tussock grass. On Hermite Island he was following in Darwin's steps, feeling that his remarks were so true and graphic that Lyell's present was 'not only indispensable but a delightful companion and guide'. Hooker's reaction to the poor Fuegians was much like Darwin's: 'the most degraded savages that I ever set eyes upon'. The climate, however, was not nearly so extreme as he had been led to believe: in their account of it Sir Joseph Banks and Dr Solander must have been reacting to the sudden change from the tropical heat of Rio – 'we thought nothing of it, and were it necessary, even without a fire, a shelter might be made, which with the warmth of two or three persons close together, might have defied death by cold'. Hooker, robust and in good health throughout the long voyage, was toughening up. The third push south began on 17 December, and on 5 January 1843 they discovered Cockburn Island, on which Hooker 'procured the ghosts of eighteen Cryptogamic plants'. The voyage into the Antarctic Circle was fraught with difficulty and danger from ice; in March the ships headed towards the Cape, with Ross and Hooker the only two reluctant to see the voyage coming to a close. Hooker was quite willing to stay for yet a sixth year, for it would give him 'a claim on the scientific world'.

Back in Britain, Hooker had more than enough material to be working up, though he lacked a well-paid permanent job which would leave him enough time to research and write. There was the *Flora Antarctica*; the *Niger Flora* which also came his way (this was the fruit of Trotter's 1841 expedition, which would probably have proved fatal to Hooker had he decided to join it); and there were Darwin's plants from the *Beagle* voyage, starting with the Galapagos Islands collection. Hooker had covered some of the same ground as Darwin, and after they met again in 1843 Darwin moved swiftly from formality and scientific collaboration, to friendship and intellectual intimacy. He urged Hooker to correlate the Fuegian flora with that of Europe, and sent him an outline of the conclusions he had formed 'regarding the distribution of plants in the southern regions'. Geographical distribution, he had decided, would be the key to 'unlock the mystery of species', and he wanted to know whether the botany of the Galapagos Islands pointed in the same direction as the zoology. It was to Hooker that Darwin first revealed his theory of transmutation.

Where Hooker was blessed with dynastic connections, Thomas Henry Huxley, Darwin's eventual champion and bulldog, had to struggle much more fiercely for his position. Not for him the Cambridge connections, the friends in high places, or Darwin's cushion of comparative wealth. Huxley's voyage on the survey ship, HMS *Rattlesnake*, a name whose briskness seems to suit him, was the making of him as a scientist: it gave him the field experience for his first papers, and his first monograph; it put a necessary distance between him and England, and the English establishment; it gave him confidence; and on its course, in Sydney, he met his future wife. But for Huxley the voyage was a career move; and he went out like Hooker, not as a gentleman naturalist, but as assistant surgeon.

Huxley was born over a butcher's shop in Ealing, the son of a schoolmaster, on 4 May 1825, and by the time he was thirteen he was working for his brother-in-law in Coventry, learning the trade of medicine. Two years later, he was apprenticed to a doctor in the East End of London, mixing drugs in the apothecary's shop in Paradise Street, Rotherhithe, where he experienced the full shock of the scarcely believable degradation and poverty of the London slums. (Alfred Wallace had seen similar sights in the late 1830s, apprenticed to his elder brother in north London, and was at this moment discovering the harsh living conditions of the agricultural poor as he carried out surveys for the railway companies in Wales and the West Country.) Huxley, seizing every chance to further his education and training, and regularly winning prizes, eventually secured a place as a free scholar at Charing Cross Hospital, and in August 1845 he passed Part 1 of the Bachelor of Medicine, with a gold medal for anatomy and physiology. In March the following year, he joined the navy, and was posted to Haslar Naval Hospital at Gosport; in December 1846 he was on the *Rattlesnake*, heading for survey tasks on the coasts of Australia and New Guinea.

Huxley was among naturalists: there was an official collector, John MacGillivray; the Stanley family's private naturalist, James Wilcox; and John Thomson, the surgeon, was also a collector, 'fond of botanical pursuits', according to Captain Stanley. Stanley himself had a wealth of experience, having honed his surveying skills off Patagonia, searched for the North-West Passage, and sailed in New Zealand waters. He made sure that Huxley received the best scientific advice available, by providing introductions to key London experts: Richard Owen, John Gray at the British Museum, and Edward Forbes of the Geological Survey. Owen advised him to concentrate on fishes' brains. (In 1857 Huxley would take furious issue with Owen on whether a man's brain was distinct from that of an ape.)

The *Rattlesnake*'s leisurely course furnished Huxley with a good personal map of the southern hemisphere: Madeira, Rio de Janeiro (where he was horrified, like Darwin, by the sight of the slaves), and Simon's Town were all ports of call on the journey to Sydney. Huxley began his scientific work by examining the sea creatures which were caught in the tow-net: first arrow-worms and sea squirts, and then, further south, Portuguese men-of-war and sea nettles. At Rio and Simonstown he dredged the harbour waters for molluscs and anemones, dissecting them under his microscope. From Simonstown, his paper on the Portuguese man-of-war was forwarded to Bishop Stanley, the captain's father, who happened to be president of the Linnean Society. As a serious scientist, Huxley was launched.

There was a stop at Mauritius, where Huxley visited the tomb

4. Huxley's drawing of Kennedy's party cutting a path through the bush, June 1848. (Archives, Imperial College, London)

of Bernardin de St Pierre's separated lovers, Paul and Virginie – even though his memory of the story was that Mlle Virginie was 'a bit of a prude' and M. Paul 'a pump'. He plucked a couple of roses from the perfect wilderness which the garden had become. Huxley was twenty-two; a moment to take stock in his journal: '"Ich kann nicht anders! Gott hilfe mir!" Morals and religion are one wild whirl to me – of them the less said the better. In the region of the intellect alone can I find free and innocent play for such faculties as I possess.' Then it was on to Van Diemen's Land and Hobart before the *Rattlesnake* eventually reached Sydney on 16 July 1847. During the three months there, Huxley prepared his material for publication. He also plunged into Sydney's social life, which he later described as a 'round of humbug – ship scrubbing, painting, calling, and being called upon – Govt. Balls and the like'; on the whole it was a dog's life. But all this changed when he met Henrietta Heathorn. She had been born in the West Indies, brought up in Kent, and emigrated with her family to Sydney, where her father owned a flour and timber mill and a brewery, and where she was helping keep house for her sister and brother-in-law. She was two months younger than Huxley. It was a rapid courtship. They danced, fell in love, and became engaged. A new era in Huxley's life began, a matter 'of much more importance than all H.M. navy put together'. The journal suffered: 'no entry for five months'.

All too soon, Huxley was at sea again, on the first of a series of four cruises which the survey expedition made: north to Moreton Bay and the Brisbane river; and then, after a brief stay at Sydney, south to the Bass Straits. The third and fourth cruises were the longest, each lasting some nine months, and the most significant. The third began in April 1848. The voyage was tedious and the pattern imposed by the survey tasks frustrating. Huxley's heart was in Sydney; his mind worried about the fate of his scientific papers, on which his future prospects rested. The weather was oppressive, with incessant rain, the ship miserable, 'hot wet and stinking'. Occasionally he came to life with some find, such as a 'singular parasite (Balanophora) upon the roots of the Gum Trees in the brush': he made a coloured drawing of it, to be sent to Hooker.

With the *Rattlesnake* sailed the *Bramble*, and the *Tam o' Shanter*, which was carrying an expedition to its starting-point at Rockingham Bay, from where Edmund Kennedy planned to travel overland to Cape York. So disillusioned was Huxley with life on board that he wanted to accept Kennedy's invitation to join the trek, but could not leave his duties for so long. However, he seized the chance of a little adventure in a preliminary reconnoitre.

Huxley enjoyed playing the part of an explorer, having a bush breakfast of damper, tea and chops, bivouacking under an opossum rug, hacking through the dense jungle scrub with a pistol in his holster and a carbine slung by his side while the 'coo-eys' of the unseen aboriginals followed their slow progress. One evening a party came into the camp, and Huxley, as he was often to do later, initiated contact: 'I went towards them and entered into conversation by jabbering and gesticulation. I fancy we were sources of great amusement to one another. I bound my handkerchief round the head of one, and obtained in return some sliced edible root wrapped up in a leaf. They invited me to their camp but I declined as Kennedy did not wish to have any close intercourse with them.' The difficulty in finding a route inland from Rockingham Bay was a bad omen. The expedition was a disaster, and Kennedy was killed within sight of Cape York. Only Jacky, the aboriginal tracker, completed the journey.

The voyage went on. Huxley was becoming very apathetic; he confessed in his journal, 'I think I never was so mortally sick of anything as of this wearisome monotonous cruise'. He distracted himself by reading: Dante's *Inferno* in Italian, not very comforting, and Goethe's *Wilhelm Meister.* He sought out meetings with the aboriginals, and sketched himself being painted by one with tribal markings across his forehead, nose and eyes. On Lizard Island, he climbed the same hill which Cook had climbed in 1770 to discover a passage which would take the *Endeavour* through the Great Barrier Reef and out to sea; the natural beauty of the scene was heightened by the recollection that 'one stood on ground rendered classical by the footsteps of the great Cook'; but then honesty intervenes, and Huxley admits that 'ought to have been heightened' is nearer the mark: 'the sun had been pouring on my back all the way up and my feelings more nearly approached sickness than sublimity when I reached the top'. The *Rattlesnake* pursued its slow course north round Cape York, on to Port Essington – '*worse than a ship*', 'about the most useless, miserable, ill-managed hole in Her Majesty's dominions' – and then past Timor and the west coast of Australia, before arriving back in Sydney in January 1849. Huxley could enjoy his leave, in the company of Henrietta, 'Ettie'; but he still had no word about the fate of his scientific papers.

The fourth cruise began on 8 May 1849; after a brief pause at Moreton Bay, the *Rattlesnake* set sail for the Louisiades, the archipelago off the east tip of New Guinea. On the initial stages Huxley was, by his own admission, a very morose animal; the weather was hot, wet, rainy, muggy, and he sat melting though half-stripped in a cabin that resembled 'an orchis-house'. Captain Stanley, almost certainly ill unknown to his crew, was another source of irritation, a mass of over-cautious hesitations and inconsistencies. But Huxley had recovered his bounce and his scientific curiosity. His scientific notebook swells again with entries on jellyfish and polyps, crustacea and molluscs, comb-jellies and salps; and he turns his enquiring, probing mind to the native peoples he encounters: Papuans, Papuo-Melanesians, and Australians. He lacked any anthropological training, for anthropology and ethnology scarcely existed; but he observed and noted, and sketched, and he spent long enough in some locations to begin to see behind the unrepresentative encounters on the beach, or on board ship, which characterised the majority of contacts. His ability at drawing provided one way to acceptance; and it was usually Huxley who took the initiative in moving beyond bartering and so gained access to a village. On Darnley Island he was even sufficiently trusted to be offered a wife.

Huxley was ready for some real exploration. The first day he saw the coast of New Guinea, 'stretching along the horizon as a blue mountainous mass', he felt that new knowledge was within their grasp: 'There lies before us a grand continent – shut out from intercourse with the civilized world, more completely than China, and as rich if not richer in things rare and strange. The wide and noble rivers open wide their mouths inviting us to enter. All that is required is coolness, judgment, perseverance, to reap a rich harvest of knowledge and perhaps of more material profit.' Also, it was necessary to take a few risks; and that Stanley refused to do: 'Cortes did not reason thus when he won Mexico for Spain, nor the noble Brooke when he conquered a province in a yacht.' Huxley had to make do with the offshore Brumer Island, where he made friends with a man he called 'the Dandy', and his two wives. They were invited to visit a village, and were so delighted by the 'primitive simplicity and kind-heartedness' of the people that they gave them three cheers on leaving, 'a proceeding which astonished them not a little'. On this beautiful island Huxley saw a glimpse of paradise, of an ideal simple life, much as Wallace would in the same area a few years on.

> The people seem happy, the means of subsistence are abundant, the air warm and balmy, they are untroubled with 'the malady of thought', and so far as I see civilization as we call it would be rather a curse than a blessing to them. I could little admire the mistaken goodness of the 'Stigginses' of Exeter Hall, who would send missionaries to these men to tell them that they will all infallibly be damned.

At Darnley Island, Huxley became friendly with a young man,

5. Huxley's drawing of himself being painted, probably at Rockingham Bay (Imperial College, London).

Do-outou, who introduced him to Kaeta, a good-looking girl who was to be his *coskeer*, his wife. This, he found, imposed all sorts of obligations on his part towards his 'wife's' relations; but it allowed him to get closer to the people, in a less artificial relationship. He pictures himself 'seated on a log with Kaeta on one side holding my pencils, on the other a little black boy who had taken an affection for me and was cuddling me most energetically, and a party of natives of all ages and sizes watching my pencil, as it delineated the outline of their houses, with vast approbation'. But the game had to be kept within limits. Huxley jokingly explained to Kaeta that as she had received so much *walli* – red cloth – on the basis that she was his wife, she should prove herself by joining the ship. When it was time to go, he found that she had been down to the boat, 'and stoutly demanded to be taken off, making use of "Tamoo's" name' as authority. 'How astonished the Sydney folks would have been at my introducing Kaeta as Mrs Huxley!' The two ways of life were not totally separate. On board the *Rattlesnake* was a Scots woman, Mrs Thompson, 'Teoma', who had been rescued from shipwreck and had lived with the Torres Straits islanders for five years. She had difficulty now in translating her native thoughts into plain English, but proved an excellent source of understanding about the local culture.

In February the ship arrived back in Sydney, and Huxley was reunited once more with Ettie, for an interlude before one final survey cruise. But on 13 March Stanley died, from an 'epileptic paralytic fit'. The *Rattlesnake* was ordered to sail for England. The history of four years, the journal written for Ettie, had come to an end.

> It tells of the wanderings of a man among all varieties of human life and character, from the ball-room among the elegancies and soft nothings of society to the hut of the savage and the grand untrodden forest. It should tell of the wider and stranger wanderings of a human soul, now proud and confident, now sunk in bitter despondency – now so raised above its own coarser nature by the influence of a pure and devoted love as to dare to feel almost worthy of being so loved.

The two had to part, perhaps for as much as three years, until Huxley was sufficiently established to be able to support a wife. The voyage took him back to Chatham via New Zealand, the Falklands and the Azores. In London he was reunited with his family, and could set about planning the publication of his zoological notes and drawings, and discovering what had happened to the scientific papers he had sent back from the other side of the world.

Huxley found, to his great encouragement, that his scientific reputation was already established. Part of his paper on the *Medusae* had been published in the *Philosophical Transactions of the Royal Society*. Professor Owen and Professor Forbes were prepared to back him. Edward Forbes, who had instructed Huxley on the use of the tow-dredge before he left, was especially warm: 'I can say without exaggeration that more important or more complete zoological researches have never been conducted during any voyage of discovery in the southern hemisphere.' With their support, he was able to secure a nominal post to the guard-ship, *Fisguard*, at Woolwich, to free him to work up his findings, while the Royal Society would back him in their publication. His research concentrated on structure, as he informed his chief and mentor, Sir John Richardson, and, as a consequence, on the reclassification of the marine animals he had been dissecting and analysing – 'more careful investigation requires the breaking-up of Cuvier's "Radiata" (which succeeded the "Vermes" [of Linnaeus] as a sort of zoological lumber-room) into several very distinct and well-defined new classes'. The position was confirmed. He had to fight, and go on fighting. Support and opportunity were there, but very little money: 'To attempt to live by any scientific pursuit is a farce,' he confessed to Ettie. 'Nothing but what is absolutely practical will go down in England. A man of science may earn great distinction, but not bread.' Perhaps he should stay in the navy, where there was at least security. Step by step, he made his mark. He was elected a Fellow of the Royal Society in 1851, at the age of twenty-six. At the British Association meeting in Ipswich that summer, a resolution was passed asking for government assistance towards the cost of publication: one of the other three names put forward with Huxley's was that of Joseph Hooker, for help in processing his Himalayan material. Huxley envied Hooker, whom he met here for the first time, so he confessed to Ettie: Hooker sat by his fiancée's side at the meeting, and was to be married any day; his father was director of Kew, and there was little doubt of his succeeding him. In addition to Hooker, Huxley met another young scientist at Ipswich, who would become a close friend and ally, John Tyndall. But no grant for Huxley was forthcoming.

Two years passed, of frustrated hopes and hard grind. He failed to be appointed to a professorship in Toronto. The tantalising possibility of a post in Sydney never materialised. He worked on. He read a paper on zebra tapeworms to the Zoological Society, with Alfred Wallace, between his Amazon and Malaysian expeditions, an admiring member of the audience. At a Geological Society meeting he met Darwin, and began to exchange papers and material with him. In 1854 things finally fell into place. He resigned from the navy rather than accept a posting to the

Illustrious; the Royal Society and the Ray Society stepped in to finance and arrange the publication of *Oceanic Hydrozoa*; and he succeeded Forbes as lecturer at the Government School of Mines in Jermyn Street. Other posts followed, including one on the Geological Survey. It was all better than Toronto. At last, he was able to write to Ettie and plan his marriage. She arrived with her family in May 1855, in very poor health; but they were too happy to postpone the wedding. 'I terminate my Baccalaureate and take my degree of M.A.trimony (isn't that atrocious?) on Saturday, July 21,' he wrote to Hooker. 'Will you come? Don't if it is a bore, but I should much like to have you there.' Hooker did go, as did John Tyndall. Darwin wrote: 'I hope your marriage will not make you idle; happiness, I fear, is not good for work.' Darwin need not have worried. From now on, Huxley always managed to combine the two; and, with Hooker, would support and complement Darwin through the next critical years. The British government may not have intended to fund the scientific revolution of the 1850s, but the opportunities provided to Darwin, Hooker and Huxley had been seized and exploited to the full.

2 *The Heart of Africa*

Darwin and Huxley stopped off at the Cape on their naturalists' journeys; Galton penetrated the Kalahari desert; but the heart of Africa presented an altogether different kind of challenge to the scientific traveller. Equatorial Africa was a place of shame, the location of one of the greatest and most inexcusable processes of extended cruelty and brutality of modern times. Even in an age used to high rates of mortality, the statistics of the slave trade's middle passage made chilling reading; and the English slave-ship *Zong*, from which in 1783 a hundred and thirty-two slaves had been thrown alive into the sea, provided just one terrible and notorious example. The continuance of slavery into an age of enlightenment, or into a more self-consciously Christian society, was unthinkable to many. An act of reparation was demanded, and it was carried out by thousands of individuals wholly committed to the cause of abolition, first, and then to attempts to redress and repair the evil which had been done to the whole structure of African society. Some of the schemes look naive and misguided in retrospect; but the philanthropic impulse was genuine and formed a part, sometimes a dominant part, of the reasons behind many early expeditions to Africa.

Africa, too, provided a site for scientific experiment. The idea that commerce, by compelling people to live more and more of their lives in public and exposing them to the moral pressures of others, could bring about change for the better lies at the heart of Jeremy Bentham's philosophy. This public life, he argued at the beginning of the nineteenth century, would necessarily make people more virtuous, and would 'continue to do so, till, if ever, their nature shall have arrived at its perfection'. Significantly, he contrasted the 'delightful abode' which could be created with 'the savage forest in which men have so long wandered'. Commerce was to be the means of bringing enlightenment, and improvement; and the commercial classes, the middle classes, would be the power-house, the energisers, of this process. This programme was translated into policy by men like Palmerston, whose approach to Africa rested on a fervent belief in the benefits of commercial

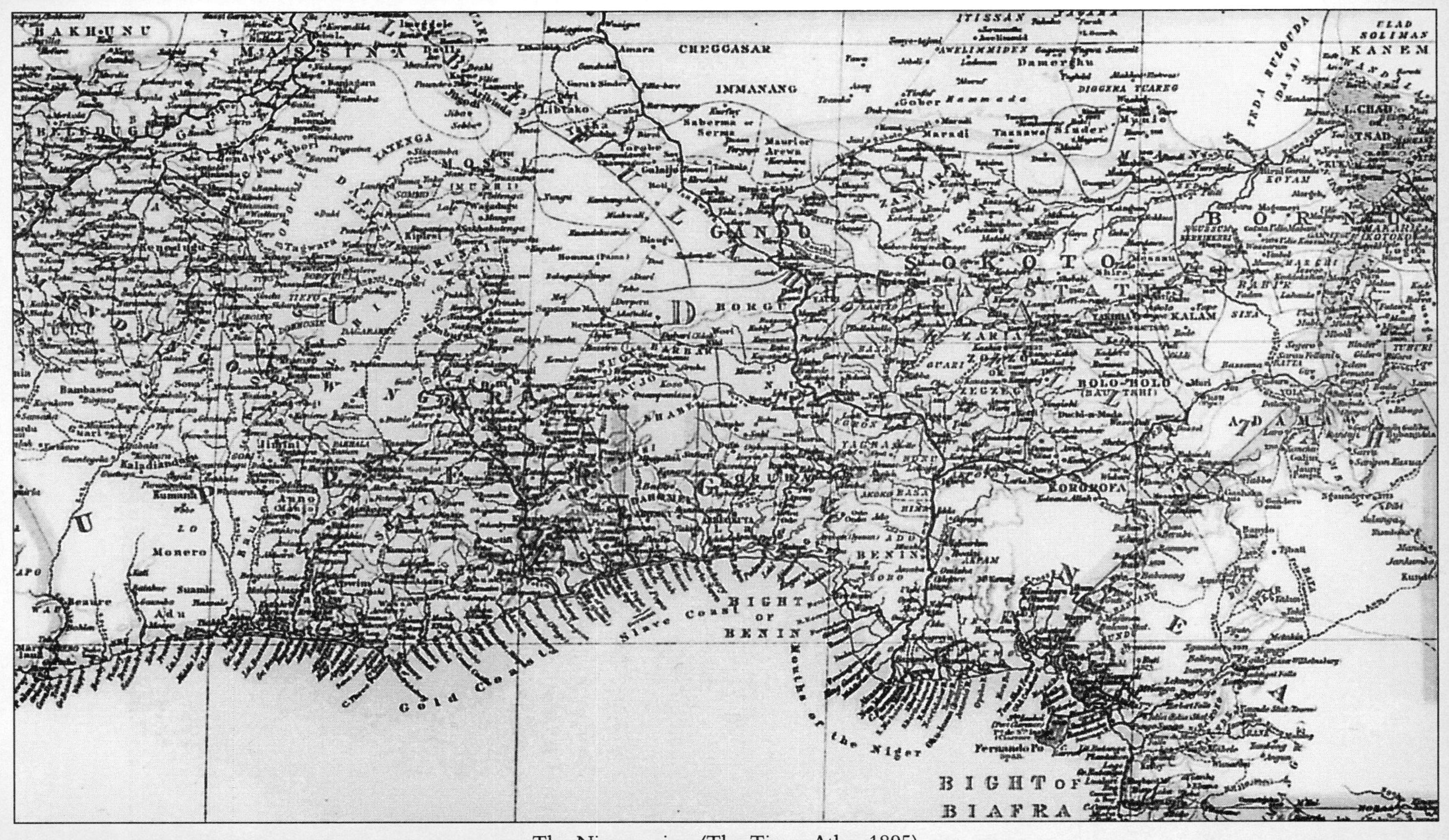

The Niger region (The Times Atlas, 1895)

expansion. It was not just a matter of self-interest, but a moral duty. The organisation of the world into different races and climates was designed to this very end, 'that commerce may go freely forth, leading civilization with one hand, and peace with the other, to render mankind happier, wiser, better. Sir, this is the dispensation of Providence. . . .' Anyone who sought to put barriers or restrictions in the way of free trade, Palmerston argued in 1842, was 'fettering the inborn energies of man, and setting up their miserable legislation instead of the great standing laws of nature'.

The early nineteenth-century explorers of Africa, swinging in their litters along the forest paths, swaying on the backs of camels across the Sahara, drifting down the rivers on canoes, were the advance troops of a great surge of optimistic commercial expansion. Part of an unstoppable process of evolution, this life-giving transfusion would spread through the trade arteries of Africa and civilise the continent: either an alternative or a complementary method to the missionary impulse.

If Africa's peoples had collectively known what the Europeans had in store for them, they might have done better to have slaughtered them all on sight.

Africa, so near to hand but so partially known, protected to some extent by the Sahara desert and the hostility of the Arabs in the north, and by disease to the south, presented a special challenge. A school atlas of 1799 shows the current state of geographical knowledge. The coastline is dotted with forts, castles and trading stations. Just inland, on the west, for example, there is a set of bald titles denoting the past trading staples: Grain Coast, Ivory Coast, Gold Coast, Slave Coast. Only the mouths of the rivers are marked, and, further inland in confident capitals, is the label NEGROLAND. The whole area of the continent is filled with names, but most of them convey no more than a naive hope, such as Lower Guinea or Lower Ethiopia; while others seem to come from medieval speculation: Desert of Seth, Anthropophagos or Men Eaters. There had been little progress since Jonathan Swift:

> So geographers, in Afric-maps,
> With savage pictures fill their gaps;
> And o'er unhabitable downs
> Place elephants for want of towns.

The Nile, the Niger – the 'Black Nile' – and the Congo were seen as ways of mapping the unknown heart of Africa; some people even speculated that they were connected (following Herodotus and Pliny), and that the Niger found its way right across the centre of the continent before turning north towards the

Mediterranean, or alternatively made a great loop and joined the Congo. South of the Sahara was the city of Timbuktu, on or close to the Niger, a name redolent of mystery and wealth; and many of the early Niger expeditions headed in this direction, either joining caravans moving south from Tripoli, or striking up the Gambia river. The African Association, formed in 1788 to promote discovery, provided a focus for the series of British sponsored expeditions which were launched towards the end of the eighteenth century. As their manifesto commented, the map of Africa was still 'a wide extended blank, on which the geographer, on the authority of Leo Africanus and of the Xeriff of Edrissi the Nubian author, has traced with hesitating hand a few names of unexplored rivers and of uncertain nations'. As no 'species of information is more ardently desired, or more generally useful, than that which improves the science of geography; so, to rescue the age "from a charge of ignorance" . . . a few individuals, strongly impressed with a conviction of the practicability and utility of thus enlarging the fund of human knowledge, have formed the plan of an Association for Promoting the discovery of the interior parts of Africa'. Among the founders was Joseph Banks, who was involved in so many scientific ventures. The 'pure' impulse to add to European knowledge went hand in hand, at least initially, with the utilitarian: the word commerce was not mentioned, but perhaps did not need to be. Humboldt became a member, as did Wilberforce. Science and the abolitionist movement lent respectability to what seems an essentially expansionist, if not yet imperial, enterprise.

The task was formidable, and the qualifications for the ideal explorer daunting: he was to be surveyor, map-maker, naturalist, anthropologist; he needed a working knowledge of a language, ideally Arabic; he should be able to draw and sketch; and have some basic medical knowledge to supplement a strong and healthy constitution. The first African Association venture was a double-headed expedition: William Lucas, who had mastered Arabic while a slave in Morocco, was allocated 'the passage of the Desert of Zahara, from Tripoli to Fezzan'; he was sportingly given a choice of return routes, by way of the Gambia or the coast of Guinea. John Ledyard, an American who had sailed with Cook on his last voyage in 1776, set off in June 1788 with an even broader brief: to penetrate into Africa by way of Egypt, and to traverse the continent from east to west in the latitude of the Niger. Lucas turned back after a few miles; Ledyard caught dysentery in Cairo, and died from an overdose of the vitriol he took to cure it. The other approach was to travel inland from the west coast, a higher risk option in terms of disease, but with the advantage of initially avoiding the hostility of the Arab traders or of tribes like the Tuareg. Major Houghton set out from the Gambia in 1791, dying before he reached his goal, Timbuktu. Four years later, Mungo Park set off 'to ascertain the course, and, if possible, the rise and terminus of the Niger River'; and, as a *de rigueur* addition, to visit Timbuktu.

In education and background, Park has some interesting parallels with men of a later generation such as Huxley. He was trained for medicine by a surgeon in Selkirk, attended lectures at Edinburgh University, and through his interest in botany was introduced to Banks. Banks's influence helped him to a post as assistant surgeon on an East Indiaman, and he sailed to Sumatra in 1792, bringing back plants for Banks. Park heard that the African Association was looking for someone to try and discover what had happened to Houghton, and, through Banks, volunteered his services. The Association judged him 'sufficiently educated in the use of Hadley's quadrant to make the necessary observations; geographer enough to trace out his path through the wilderness; and not unacquainted with natural history'. Park's stated aims provide a clear picture of the prevailing attitude to Africa:

> I had a passionate desire to examine into the productions of a country so little known, and to become experimentally acquainted with the modes of life and character of the natives. I knew that I was able to bear fatigue; and I relied on my youth and the strength of my constitution to preserve me from the effects of the climate. The salary which the Committee allowed was sufficiently large, and I made no stipulation for future reward. If I should perish in my journey, I was willing that my hopes and expectations should perish with me; and if I should succeed in rendering the geography of Africa more familiar to my countrymen, and in opening to their ambition and industry new sources of wealth, and new channels of commerce, I knew that I was in the hands of men of honour.

Park sensibly began his first expedition with a six-month acclimatisation period on the Gambia, studying the key language, Mandingo, and learning all he could there from one of the European residents, Dr Laidley: it is a sign of the time that Laidley was a slave trader, not a medical missionary. At the beginning of December Park traced his way up the Gambia river, and over a tributary of the Senegal to Jarra, where he found Houghton's remains. He survived a period of virtual imprisonment in an area he described as a land of thieves and murderers, during which he

6. Mungo Park in despair after being robbed. (Title page, *Travels in the Interior Districts of Africa*, Edinburgh, 1860)

was often sustained by the kindness of women who took pity on him. (They sang as they spun cotton: 'The poor white man, faint and weary, came and sat under our tree. He has no mother to bring him milk; no wife to grind his corn. Chorus: Let us pity the white man; no mother has he, etc.') Travelling south-east, Park came to the Niger, flowing, significantly, west to east, and followed it downstream for three hundred miles or so. He was the first British traveller actually to see the upper course of the Niger, and return to tell the story.

The story was thrillingly told. *Travels in the Interior Districts of Africa* came out in April 1799, and earned Park 1,000 guineas. It was a landmark in European travel writing, and Park's description of his first sight of the Niger placed the river firmly within a British context: 'the long sought for majestic Niger, glittering to the morning sun, as broad as the Thames at Westminster, and flowing slowly to the eastward'. The African Association rejoiced. The value of Park's discovery was self-evident: 'a gate is opened to every commercial nation to enter and trade from the west to the eastern extremity of Africa'. But if Park had proved that the Niger flowed from west to east, it was still unclear where it came out. The search for the precise course continued.

Park settled down to write up his travels, to marry, and to go into medical practice in Peebles. The lure of Africa remained; to friends who reminded him of the dangers, he would reply that a few long Border winters were just as much of a threat to life. Not long after the Peace of Amiens was signed in 1801, Sir Joseph Banks was in touch once more. The Association was certain to revive their project of a mission to penetrate to and navigate the Niger, and 'in case Government should enter into the plan, Park would certainly be recommended as the person proper to be employed for carrying it into execution'. Government arrangements moved more ponderously than the Association's; Park prepared himself by practising with astronomical instruments, and studying Arabic. Walter Scott found him one day on the banks of the Yarrow, plunging stones into the river and watching as the bubbles rose to the surface – Park explained that this was his well-tried method of judging the depth of an African river. At last, the summons came.

Park's second journey in 1805 took place on a very different scale: thirty soldiers, six seamen, all Europeans, £5,000, and an ambitious plan to retrace his route, build two boats, and follow the course of the river to one of its two supposed destinations, either the lake or swamp known as Wangara, or the Congo system. When Park reached Bambakoo on 19 August, where he '*once more saw the Niger* rolling its immense stream along the plain', he had only seven Europeans left alive out of his original party. By the time the boats had been built, they were down to five. Park sent back journals and letters to the coast with one of his African guides, Isaaco: 'Though all the Europeans who are with me should die, and though I were myself half dead, I would still persevere, and if I could not succeed in the object of my journey, I would at least die in the Niger.' Richard Lander, following the same route years later, was informed that Park's party had been mistaken for the heavily armed war party of a neighbouring tribe; his boat was attacked, and Park drowned.

The renewed Napoleonic wars in Europe postponed further government initiatives, but following the peace a succession of attempts, led by army or naval officers, tried unsuccessfully to solve the question of the Niger's course. Captain Tuckey died on the Congo; Major Peddie, with a hundred men, was wiped out on the Nunez – two prongs of a pincer movement designed to meet each other somewhere on the Niger. Joseph Ritchie and Captain Lyon, travelling as Muslims, penetrated about six hundred miles south of Tripoli; Ritchie died, Lyon barely escaped with his life. Between 1822 and 1824, Major Dixon Denham, Dr Walter Oudney and Captain Hugh Clapperton successfully reached Bornu, again travelling south from Tripoli; while Denham explored the Lake Chad area, Clapperton visited Kano and Sokoto, and brought back a letter for King George IV from the Fulani ruler, Sultan Muhammad Bello. Bello offered to prohibit the export of slaves, and indicated that, in return for supplies of arms, he would allow a British consul to be posted in his territory. This was progress. Clapperton was duly launched on a fresh expedition, this time aiming north in 1825 from the coast at Badagry, near Lagos. Clapperton, adventurous and spirited, did succeed in reaching the Niger, but the only person who survived the trip was his servant, Richard Lander, who returned after overcoming much difficulty to Badagry, the starting-point.

Richard Lander, as the introduction to his *Journal of an Expedition to Explore the Course and Termination of the Niger* is at pains to point out, was a very different kind of individual to the medical men or the experienced naval and military officers who had led previous expeditions. For a start, he had little formal education. His background, and the way he was shaped by experience, makes him one of the most unusual personalities of this phase of African exploration. He was a Cornishman, born in 1804, and his father kept a public house, the Fighting Cocks, in Truro. At the age of eleven he was taken out to St Domingo by a West Indian merchant, presumably as a servant. Ill with fever, he recovered, 'owing chiefly to the kindness and attention I experienced from some benevolent and sympathizing Negro females'. This exposure to the tropics and to African society seems to have stayed with him vividly, as a key factor in his later and distinctly non-standard approach to Africa and Africans. He returned to England, entered the service 'of various noblemen and gentlemen' in France and elsewhere on the continent, and then accompanied Major Colebrook on his Commission of Inquiry into the state of South Africa. Lander was back in England when Clapperton and Denham returned from their expedition, and when he learned that Clapperton was going to return to Africa on a British government expedition to explore 'the yet undiscovered parts of central Africa, and of endeavouring to ascertain the source, progress, and termination, of the mysterious Niger', he volunteered, and was engaged as Clapperton's 'confidential servant'.

Lander's motives are hard to unravel. He knew very clearly the

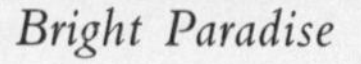

7. Richard Lander. (National Portrait Gallery)

dangers of the mission, and his friends and family made repeated efforts to dissuade him from going. A family connection even promised him the offer of a 'more lucrative situation in one of the South American republics' as an alternative. But Lander had been changed by his travels. His retrospective explanation was blatantly romantic: 'There was a charm in the very sound of Africa, that always made my heart flutter on hearing it mentioned: whilst its boundless deserts of sand; the awful obscurity in which many of the interior regions were enveloped; the strange and wild aspect of countries that had never been trodden by the foot of a European, and even the very failure of all former undertakings to explore its hidden wonders, united to strengthen the determination I had come to.' Africa represented a special challenge, and offered him the kind of opportunity for which his previous experience had helped to prepare him.

In terms of its aims, the expedition was a disaster, as some of the illustrations to Richard Lander's 1830 account, *Records of Captain Clapperton's Last Expedition to Africa*, confirm: 'Funeral of Captain Clapperton'; 'Fainting Scene'; 'Ordeal by Poison'. From their base at Badagry, they began a laborious set of negotiations to enable them to move into the interior. By page 35 of Volume 1, Dr Dickson, a Scottish surgeon, who had struck out on a different route, had died, the first 'victim to the cause of African research'. The following chapters contain a bleak record of deathbeds and funerals, as Dawson, a seaman, the naval captain Pearce (the potential consul), and Dr Morrison succumbed in turn to fever or dysentery. Clapperton himself was frequently ill – once, interestingly, cured by native medicine – and Lander had several bad bouts of fever, but the two survived, and continued north. In places, the journal reads as a kind of pilgrim's progress, a set of tests and encounters sent to try the faith of the survivors. This tendency becomes even more marked when the narrative reaches the Muslim areas, where the white man becomes defined as 'Christian'. Elsewhere, it is a romantic odyssey, enlivened by exotic African entertainments or the occasional glimpse of Eden:

> At noon we descended into a delightful valley, situated in the bosom of a ridge of rocks, which effectually hid it from observation till one approached almost close to it. It was intersected with whimpering [*sic*] streams and purling rills, and the broad leaved banana, covered with foliage, embellishing the sheltered and beautifully romantic spot. In the centre was a sheet of water, resembling an artificial pond, in which were numbers of young maidens from the neighbouring town of Tschow, some of them reposing at full length on its verdant banks, and some frisking and basking in the sun-beams, whilst others of their companions were sporting with the Naiads of the sacred stream; but all of them visibly delighted with the pleasant recreations which they were enjoying so prettily and innocently.

This scene of innocence, heavily dependent on the stock motifs of neoclassical painting, could be chosen today as a typical example of the literature of oppression: the male, white gaze intruding, voyeuristically, on the female Africa. But Lander's account shows some awareness of the intrusive presence. As soon as the 'white faces' were observed, he continued, the young ladies ran away and hid behind the trunks of trees, 'looking as coy and bashful as did their mother Eve in the garden of Eden'. To locate Eden and innocence in the heart of Africa, on a journey of this kind, was a comparative rarity at this date. Ideas of paradise were attached to the islands of the South Pacific, rather than to equatorial Africa. Mungo Park provided a reasonably sympathetic account of the Mandingo people, but his attitude was inevitably coloured by the period of imprisonment he underwent and his perspective was more systematic and anthropological: he saw the drawbacks, and the diseases. The encounters he describes with women are either of the charity he received as a destitute pilgrim, or of the 'lucky escape' variety – he recounts how he extricated himself from a visit by a group of 'Moorish ladies' who called at his hut 'to ascertain, by actual inspection, whether the rite of circumcision extended to the Nazarenes (Christians), as well as to the followers of Mahomet'. Major Denham adopted a breezy, racy style in describing his encounters with the beautiful girls of Bornu, who introduced him to the custom of 'shampooing' – a back massage with oil or fat: 'Verily I began to think that I not only deserved to be a sultan, but that I had really commenced my reign . . . ' Lander's accounts are unusual in their enjoyment and appreciation of many aspects of African culture. Part of this may spring from his own early experience of kindness from African women; part from his own nature. It becomes clear that he was a good-natured, extrovert, short, plump, happy young man – he was only twenty when the expedition began – as well as having a strong constitution. He was also used to hard work, and had a kind of innocence himself. As the two men, the Scottish army captain and his Cornish manservant, continued their journey, the barriers of class and education began to crumble. Seen by African eyes, the whiteness of the two men was a stronger unifying factor than any other. But Clapperton decided that it would be safer to change the nature of the relationship, and introduced Lander as his son rather than his servant.

Crossing over the Niger, they continued to Zaria and Kano, where Clapperton left Lander with the baggage while he went to Sokoto alone: he needed permission to travel through Bornu, and so back across the Sahara to Tripoli, as he did not wish to make the journey down the Niger. By this time Lander, even in letters, was 'Richard'. Before long, he was summoned to join Clapperton in Sokoto. They spent a peaceful, if frustrating, few weeks, shooting during the day, and in the evenings smoking cigars and talking about the past. Lander had his buglehorn, and would play 'Sweet, sweet home', or sing 'My native Highland home' to Clapperton. Then Clapperton fell ill with dysentery. Lander nursed him through his final illness, received his last instructions, and in due course buried him: 'uncovering my head, and opening a prayer-book, I read the impressive funeral service of the Church of England over the remains of my English master – the English flag waving slowly and mournfully over them at the same moment'. (When Captain Lyon emerged from the desert near Tripoli in 1820, he chanted 'God Save the King' and 'Rule Britannia' as loud as he could roar at the sight of the Mediterranean.)

Lander was in despair, and ill with fever: 'I felt as if I stood alone in the world, and wished, ardently wished, I had been enjoying the same deep, undisturbed, cold sleep as my master, and in the same grave.' Lander had a Hausa servant with him, called Pasko, and he had picked up enough Hausa to make himself understood. But he was a hundred days' journey from the coast. He did not have the advantage of the contacts with the chiefs and emirs which Clapperton had established on his previous journey; and he had little inclination to make the desert journey that Clapperton had recommended. In fact, Lander felt far more at ease with the Negro races than with the Arabs. He still wanted to try the Niger route, and he struck off south-west, imagining that he could float down the river by canoe to Benin, but was overtaken by armed horsemen and had to retrace his route. He headed this time towards the Yoruba kingdom, taking with him Pasko, and Aboudah, a girl presented to him as a wife (he rather skates over this relationship in his book), together with Clapperton's 'papers', the preservation of which had become by this time his sacred mission.

Lander pads out the story with a lively but unsystematic account of peoples and their customs. When he arrived at Badagry, he had one final ordeal to face. He was denounced, he guessed, by Portuguese slave traders and forced to drink a 'poison chalice'. 'You are accused, white man, of designs against our king and his government, and are therefore desired to drink the contents of this vessel, which, if the reports to your prejudice be true, will surely destroy you; whereas, if they be without foundation, you need not fear, Christian; the fetish will do you no injury, for our gods will do that which is right.' Lander offered up a short prayer to the God of Mercy, and swallowed the bowlful. To everyone's amazement, he appeared to suffer no ill-effects. However, he was taking no chances: as soon as he got back to his hut, he dosed himself with an emetic and made sure he vomited. All was now well as far as his safety was concerned; but it still took two months before an British ship arrived in the vicinity and he could arrange a passage to Cape Coast. Lander quickly became very British again. 'My heart bounded within me at sight of the proud banner of my country, streaming from the stately mast of the Maria . . . and as the little bark moved swiftly towards the ship, although my appearance was pitiable in the extreme, the yards were instantly ordered to be manned, and three tremendous cheers from the throats of British seamen, welcomed me once more to the dear society of my gallant countrymen.' Lander's status had altered. He was received by the Governor of Cape Coast, and dined by the British merchants. He freed his slaves, who were given money and a plot of ground by the Governor. He then headed home on a naval sloop, via Fernando Po, St Helena, and Ascension islands.

Lander included a curious incident which, more than anything, helps to explain the hold that his vision of Africa exercised over him. Some turtles had been caught at Ascension, and Lander was asked by the captain to bathe their eyes morning and evening, in an attempt to keep them alive. Lander recoiled. 'Could I, who had so recently shaken hands with majesty, and lodged in the palaces of kings; who had been waited on by the queen of Boussa, and walked delighted with the proud princess of Nyffe . . .' and on and on – 'could I so suddenly fall from the towering elevation to which my consequence had raised me, forget all my dignity and all my laurels, and descend to the menial, grovelling occupation of washing the filthy eyes of turtles?' The man who had been hailed as 'Nassarah Curramee' in Africa, the god, the prophet, the enchanter, had vanished, and degenerated, in the eyes of his fellow Europeans, into the country boy and servant of before. As a passenger, he declined the task. He was honest enough to report both his feelings and the fact that the turtles died. He dutifully delivered all of Clapperton's effects to the Colonial Office, including, he took care to tell his readers, gold and silver watches. He seems to have received very little praise or gratitude for his efforts, and when his account was first published he was criticised for sounding much too familiar in his description of his master, Clapperton.

The simple Richard Lander who had set out as an anonymous servant had, at least in his own eyes, been transformed – into someone who could write and publish his autobiography, for that is what his second account of the journey becomes. The first he had written to 'satisfy Ministers' with regard to his conduct 'after the decease of Captain Clapperton . . . and to make them acquainted with the manner in which the property left in my charge at Soccatoo had been disposed of'. In the second he included 'a thousand amusing incidents which had been overlooked', and attempted to depict the customs and ceremonies of the nations and tribes between Badagry and the Hausa kingdom. The serious scientific traveller John Barrow, reading the book for the publisher John Murray, criticised Lander for his style and for 'sins of egotism'; the sketch of Lander's life was 'utterly unimportant and uninteresting', while the work as a whole was 'deplorably meagre in notices connected with the botany, zoology etc.'

That last accusation is true enough; Lander's interest was in people, himself included, which is what makes his book so readable. The journey he narrates in *Clapperton's Last Expedition* is not only a modern-day *Pilgrim's Progress*; it is an account of a servant who was faithful to the end; a picaresque novel, which returns the hero safely to his friends in Truro after an absence of thirteen years – so that the adventure, for Lander, begins when he first sets off from home to travel to the West Indies; a *Bildungsroman*. But it also left Lander with an unsatisfied curiosity, a sense of unfinished business. He had twice crossed the Niger, yet knew no more about its course than before. General Sir Rufane Donkin published an extravagant theory which suggested that the Niger passed through Lake Chad, went underground for a thousand miles, and emerged into the Mediterranean somewhere in the Gulf of Sirte. Lander thought little of this. In December 1829, after completing his book on Clapperton, he volunteered to return to West Africa to complete his mission, by tracing the Niger from Fundah to Benin. Barrow, a civil servant at the Admiralty, intervened. Barrow had been asked for his comments by Hay, Under-Secretary at the Colonial Office, on two independent plans to solve the Niger question. 'Would not Lander, who has been pressing to go again, be the fittest person to send? No one in my opinion would make their way so well and with a bundle of beads and bafts and other trinkets, we could land him somewhere about Bonny and let him find his way.' Lander, unregarded as a writer, and clearly quite expendable, had as good a chance as anyone. (Both the other independent explorers who set out at this time died at their journey's beginning.) As a special request, Richard's brother John was allowed to go with him, without a salary. The Landers' plan is worth quoting, for it shows a sensitivity and a shrewdness largely absent from the first wave of African explorers: 'We shall endeavour to conform ourselves, as nearly as possible, to the manners and habits of the natives; we will not mock their blind superstition, but respect it; we will not scoff at their institutions, but bow to them; we will not condemn their prejudices, but pity them. . . . Confidence in ourselves, and in them, will be our best panoply; and an English Testament our safest fetish.' (This was a distinctly naive view of the Landers' status as Christians, once they moved away from the coastal districts; most of the travellers who passed through the Arab sphere of influence took good care to pass themselves off as Muslims.) If they should perish in Africa, Lander added in an ironic last sentence which reflects the cool disdain with which he had been treated, it was some consolation to know 'that the gap we may make in society will scarcely be observed at all'.

Lander was promised a pension, and a pension for his wife, *after* the expedition: his brother John was to receive nothing. He was given a free passage to Badagry, $200 for travelling expenses, and extremely basic equipment together with a medicine chest from the stores of the London military depot. The government must have felt that they were getting him on the cheap. Lander's orders were set out in wonderfully vague terms:

> If you should find that at Funda the Quorra [Niger] continues to flow to the southward, you are to follow it to the sea . . . but if it should be found to turn off to the eastward, in which case it will most probably fall into the Lake Tshad, you are to follow its course in that direction, as far as you can conceive you can venture to do, with due regard to your personal safety, even to Bornou, in which case it will be for you to determine whether it may not be advisable to return home by the way of Fezzan and Tripoli.

Lander was also ordered to take every opportunity to send down by the coast 'a brief abstract of your proceedings and observations, furnishing the bearer with a note, setting forth the reward he is to have for his trouble, and requesting any English person, to whom it is presented, to pay that reward, on the faith that it will be repaid him by the British Government'. There was no need to spell out the reason for this request for interim reports.

Lander's definitive expedition in 1830 provided the crucial breakthrough in knowledge of the topography, and of the lines of communication, of West Africa. It was very different from most scientific enterprises, more in the nature of a suicide mission: put the young men ashore, and let them find their way. From Richard Lander's point of view, there was a serious geographical purpose, and his enterprise seems infinitely more purposeful than, for example, Francis Galton's trek in southern Africa in 1850, for which he was awarded the Royal Geographical Society's gold medal. Lander went to solve a problem, even if he went equipped with the minimum of instruments: two compasses, two thermometers, and one watch (common silver), all supplied from a military depot: 'Delivered out of His Majesty's Stores at this place, by an Order of the Honourable Board of Ordnance . . . to Messrs. Lander, about to proceed on Discovery in Africa.' The rest of the stores comprised trade-goods: cloth, mirrors, scissors, knives, combs, beads, and 50,000 needles – 'Whitechapel sharps' – as well as basic camping equipment – 'Tent, circular, complete, 1, Ditto, pins, 40, Mallets, tent, 2' – and even writing materials: 'Ink bottle, small, 1, Books, journal, thick quarto, 2, Ditto, memorandum, 2.' When the Landers arrived at Cape Coast Castle, they were additionally provided with forty muskets and twelve signal rockets as a present for the king of Badagry. Guns, trinkets and rum were the instruments of progress.

Perhaps the most significant items of equipment were two cases of medical supplies, with detailed instructions for use supplied by Sir John Webb, including four bottles of sulphate of quinine. Quinine pills were to be taken 'as a strengthener after fever or dysentery'. The effectiveness of quinine in preventing malaria would not be established until the Niger trip of 1842, although the remedy had been known and used for years. Purgatives, blisters and bleeding were the ferocious treatments applied to fever, though Lander may have taken enough quinine when recovering to act as a partial prophylactic against the next attack.

The Landers made their way along the coast, from fort to fort and trading station to trading station, until they were finally put ashore at Badagry. By this time they had employed old Pasko, and his wife Aboudah (who had originally been presented as a slave to Richard in Kano); another former slave, Jowdie; two men from Bornu, Ibrahim and Mina; and Antonio, son of a Bonny chief, who thought with encouraging optimism that he would be able to reach home by way of the Niger. The first part of their journey was fraught with difficulties and frustrations. Their account of this first leg of the expedition is, largely, from John's journal, since Richard's notes were lost in the Niger. John was clearly less tolerant than his older brother, less good-humoured, less inquisitive; he was better educated, with an elaborate prose style to match, full of quotations from English poetry. He was also more prone to illness, which no doubt affected his buoyancy. The negotiations for progress were protracted. King Adooley wanted a great deal in return for his protection on the next stage: his first shopping list included two long brass guns, fifty muskets, twenty barrels of gunpowder, half a dozen rockets and a rocket gun; the next day, a gunboat and a hundred men from England were added to the request, and the letter sent to Cape Coast. Lander was becoming worried that he would not have enough presents in his luggage to last the rest of the journey. He was already struck by a marked difference in reception compared to his previous expedition. When he reached Katanga – old Oyo, the Yoruba capital – Mansolah, King Majotu, appeared ungracious and ungenerous: 'The king has sent us nothing since the day of our arrival; and the present then given was disgraceful in the extreme, coming from the monarch of a large and mighty kingdom.' On Lander's previous visit with Clapperton five years before, the white men were believed to be 'messengers of peace', 'charged with a commission to make peace wherever there is war and to do good to every country'. Lander now felt that the king and his subjects had seen quite enough of white men; 'the rapturous exultation which glowed in the cheeks of the first Europeans that visited this country on being gazed at, admired, caressed, and almost worshipped as a god – joined to the delightful consciousness of his own immeasurable superiority, will, in the present age at least, never be experienced by any other. Alas! what a misfortune! The eager curiosity of the natives has been glutted by satiety – an European is shamefully considered to be no more than a man!' The satiety, of course, was two – no other Europeans had passed this way since Lander himself. One white expedition was a curiosity, to be absorbed and accommodated. A second was more ominous, and spelt change. No more goat and mutton – only grilled rat and roasted locust.

The Landers made their way to Bussa, on the Niger, where Richard had been received with great kindness in 1826 by the ruler and his wife. It was also the scene of his encounters with the exuberant and vast widow Zuma, who had pressed herself on him as a suitable wife, before turning her attention to Captain Clapperton, equally in vain, as long-stop. This time, although they were generally well received, a note of frustration appears in the accounts of their stay – not surprisingly, perhaps, because it took three months before they received permission to set out on the real object of their journey, much of the time being used up in endless negotiations for suitable canoes. All the while, their supplies of goods were disappearing, and the diversions that took place – horse-racing on the Muhammadan Sabbath, the eclipse of the moon – were only partial compensations, though they certainly add colour to the story. The canoe problem was solved in two stages; the Landers bought a second at Rabba, having sold a quantity of needles in exchange for cowries, and took possession of one 'of an immense size' – fifteen feet long and four wide, which does

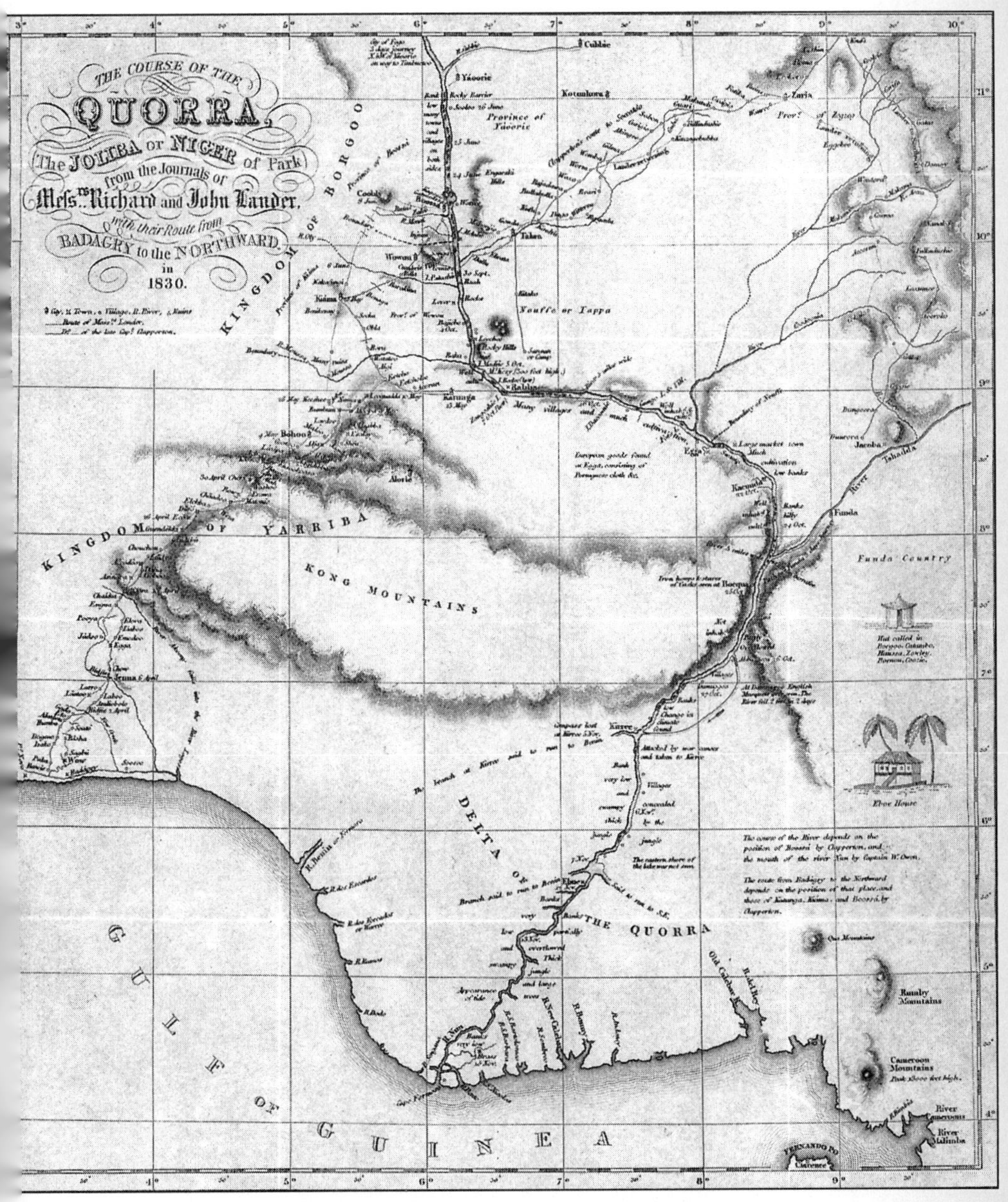

The Landers' map, showing the course of the Niger, 1830

not seem much to go down the Niger in. No one, however, would sell them any paddles, so they had to steal some. On 16 October, at nine o'clock in the morning, they 'bade farewell to the King of the Dark Water and the hundreds of spectators who were gazing' at them, fired two muskets, and gave three cheers. Launching out into the river, they were soon out of sight.

The Landers were leaving a highly organised, industrious area for the unknown. The Nupe, among whom they had been staying on the south bank of the Niger, were in control of the local river trade. The chief, they calculated, had 600 canoes. The Nupe farmed the land around; more surprisingly to the European eye, they produced excellent cloth, not at all inferior to Manchester goods, and all sorts of other items. 'In our walks we see groups of people employed in spinning cotton and silk; others in making wooden bowls and dishes, mats of various patterns, shoes, sandals, cotton dresses and caps, and the like; others busily occupied in fashioning brass and iron stirrups, bits for bridles, hoes, chains, fetters, etc.; and others again employed in making saddles and horse accoutrements.' This section of what is now northern Nigeria formed an almost self-sufficient economy; as the river flowed on, first south-east and then south, the pattern of life became far more complex, and difficult to interpret. They met the usual hazards of river travel: a herd of hippopotamus, a giant crocodile. But at each town or village they stopped at they were made to feel apprehensive, as the inhabitants warned them about the dangers to come. The Landers spread anxiety among their crew – they still had with them the Hausa Pasko, Jowdie, Antonio from Bonny, and a Sierra Leonean – by offering up prayers to the Almighty Disposer of all human events 'that he would deign to extend to us his all-saving power among the lawless barbarians'. On 25 October, they found themselves opposite a great river entering the Niger from the east, three or four miles wide at its mouth. They assumed this to be the 'Tshadda', flowing from Lake Chad. It was the Benue, 'mother of waters', the great tributary of the Niger, and the Landers were the first Europeans to see it.

Later that day, the Landers had an encounter with the local people which helps to explain their success. They had been encouraged by finding, at their proposed camping site, three or four staves of a powder barrel, and assumed that trade in gunpowder meant contact with Europeans on the coast. They were settling down in their camp when some of their men appeared, to warn them that a huge party was on its way, armed with muskets, bows and arrows, knives and spears. Pasko and the rest were given muskets and pistols, but told not to fire unless they were fired on first.

The two brothers threw down their own pistols and walked composedly towards the chief. 'His quiver was dangling at his side, his bow was bent, and an arrow, which was pointed at our breasts, already trembled on the string.' Just as the chief was about to pull the cord, a man near him stayed his arm. 'At that instant we stood before him, and immediately held forth our hands; all of them trembled like aspen leaves; the chief looked up full in our faces, kneeling on the ground – light seemed to flash from his dark, rolling eyes – his body was convulsed all over, as though he were enduring the utmost torture, and with a timorous, yet undefinable, expression . . . he drooped his head, eagerly grasped our proffered hands, and burst into tears.' Peace and friendship triumphed. Later, after presents of needles, and having found an old man who spoke Hausa to interpret, the chief expanded on the moment. He had, very reasonably, assumed that his village was threatened. But when he saw the white faces, and when Lander drew near and extended his hand, he believed the two Europeans were 'Children of Heaven', and had dropped from the skies. Presents of yams and kola nuts followed, and the welcome information that after another seven days' travel they should reach the sea. They moved on early the next morning, paddling through the silent forest: 'the song of birds was not heard, nor could any animal whatever be seen: the banks seemed to be entirely deserted, and the magnificent Niger to be slumbering in his own grandeur'.

The idyll did not last. Near Asaba they were confronted by a fleet of 'war' canoes, and attacked. Though shots were fired and blows struck and one canoe sunk, no one, amazingly, was injured on either side. The aggressors came from further downriver, and the Landers were rescued by the people of the nearest village; complicated negotiations ensued, and some of the waterlogged articles were recovered. But the Landers' independence had gone, along with treasures like the nine ivory tusks they had been given, Mungo Park's double-barrelled gun, and their remaining stock of cowries and needles. They were defenceless, and without means. Neither the Landers nor their followers were harmed, but they had, effectively, become commodities, and the question was how much they were worth. Antonio, so close to his home in Bonny, was pleading their cause, but was outnumbered by the entrepreneurial men from Brass. 'We have all been to the chief,' Lander reported Antonio as telling him, 'crying to him, and telling him that black man cannot sell white man; but he will not listen to us; he said he would sell you to the Brass people.' The white man has been reduced to the status of slave: this is the lowest depth which Lander can conceive, far more painful than his isolation when

Clapperton died, or when he lay apparently within hours of death helpless with fever. 'In most African towns and villages we have been regarded as demi-gods, and treated in consequence with universal kindness, civility, and veneration; but here, alas! what a contrast, – we are classed with the most degraded and despicable of mankind, and are become slaves in a land of ignorance and barbarism, whose savage natives have treated us with brutality and contempt . . . the consciousness of our own insignificance sadly militates against every idea of self-love and self-importance, and teaches us a plain and useful moral lesson!' Lander's sense of himself as a Christian pilgrim, being tested to the utmost, came to the foreground in this final phase of his journey, sustaining him through the bewildering chaos of his virtual captivity.

The eventual plan which emerged from protracted negotiations was to find an English ship and bargain with the master for an exchange; the Landers' value was set at twenty slaves, plus an extra charge of fifteen slaves, or casks of palm oil, for the trouble of transporting the two men down to the sea: the total sum was recalculated as ten muskets. The Landers were ferried down to Brass in the custody of Chief Boy, and understandably present a slightly jaundiced view of the place and its inhabitants. Richard was separated from his brother and taken by canoe down the river, where he saw two vessels at anchor, a Spanish slave-vessel, and a British brig. His initial feelings of delight soon changed. Four of the British crew had just died of fever; four more were lying sick in their hammocks, and the captain himself was convalescing from a near-fatal bout. Lander produced his official papers from the Secretary of State and the Lords of the Admiralty, and duly explained his situation. He had the authority to name a reward, which he proposed should be paid in the form of twenty trade guns, and the written confirmation of Downing Street that the value of the goods would be repaid. Lander breathed a sigh of relief: the long arm of British imperial rule was about to come into action.

He was entirely mistaken. Captain Lake turned out to be an archetypal palm-oil ruffian, and responded with 'the most offensive and shameful oaths'. 'If you think that you have a bloody fool to deal with, you are mistaken; I'll not give a bloody flint for your bill.' Lander was shattered. His brother, and eight people dependent on him, were effectively hostages in Brass; and he had given his personal word that the 'bill' would be honoured. With great difficulty, Boy was persuaded to go back to Brass, and bring John Lander and the rest to the brig, where the debt, they were promised, would be paid. Lake was waiting for a favourable wind, so that he could make off across the bar and out of the river without paying a pilot, and was obviously quite prepared to sail at any moment if the opportunity arose, abandoning not only Lander but his own first mate, who had left the safety of the ship to test the depth of the bar, and been captured. John Lander was finally got on board along with the mate, but only after Lake had threatened to bring 1,000 men-of-war to burn down the town and kill all the local inhabitants. Lander left with considerable feelings of guilt about the way Boy had been bullied, and the way he had been forced to break his word. (The British government later took action to ensure that the debt was paid, since this kind of bad faith could undermine trade relationships for years.) Lake took the Landers to Fernando Po, where they were extremely relieved to see the end of him: literally, for when he eventually left port, they watched like spectators at a nautical melodrama while another ship 'with long raking masts' appeared from behind a different part of the island and proceeded to open fire, and grapple with the brig. Outpirated, Lake and the brig *Thomas* were never heard of again.

The Landers made a number of short voyages to Calabar, and finally left Fernando Po and Africa for Rio de Janeiro at the end of January. The death rate continued even on that part of the voyage. Half the crew died, victims of fever, and one Krooman, 'Yellow Will', fell in the sea and could not be rescued: they heard his cries 'nearly an hour afterwards' with the most painful feelings. They finally reached Portsmouth on 9 June 1831, eighteen months after they set out.

Lander's achievement, in twice penetrating to the central reaches of the Niger and twice returning alive, and of completing the work of Mungo Park by tracing the river's course down to the sea, was outstanding. He had done it with the minimum of cost, and perhaps his most remarkable skill lay in his ability to negotiate himself out of a tight corner. The *Edinburgh Review* described the voyage as 'perhaps the most important discovery of the present age'. The way into the heart of Africa was revealed; no longer would it be necessary to trek south across the Sahara, or to push inland through the Yoruba country, dependent in each case on the goodwill of a succession of chiefs and rulers. Yet Lander's initial rewards, from the government at least, were modest. His contract had been for £100, and that is what he got. His brother, promised nothing, received nothing. But they were paid 1,000 guineas advance from John Murray, who had rejected Richard's narrative of Clapperton's last journey, for their journals of this expedition; and Richard received 50 guineas and a citation from the Royal

Geographical Society, the first of a long line of distinguished gold medallists. This was both a recognition of his achievement, and a sign of the potential importance of the Niger route. The introduction to the *Journal*, however, makes it quite clear what the establishment thought of his map-making: 'The accomplished surveyor will look in vain along the list of the articles, with which the travellers were supplied, for the instruments of his calling; and the man of science, to form his opinion of it, need only be told, that a common compass was all they possessed to benefit geography, beyond the observation of their senses. . . . Too much faith must not therefore be reposed in the various serpentine courses of the river on the map, as it is neither warranted by the resources, nor the ability of the travellers.' But the sketch would assist future travellers, from whose superior attainments 'something nearer to geographical precision' might be expected.

Richard Lander's horizons, though, had changed. He had become absorbed by a vision of Africa, in which dreams of wealth and status prevailed. Alongside his vivid appreciation of social customs came a shrewd appreciation of the Niger as an aid to trade. Even the map he drew, in the absence of a compass, marked the pattern of commerce: 'European goods found at Egga, consisting of Portuguese cloth etc.'; 'At Damuggoo English musquets were seen.' The Liverpool merchants sensed the potential of the Niger as a trading route leading directly into the heart of West Africa, bypassing the unhealthy and volatile coastal zone. The African Inland Commercial Company was founded, headed by Macgregor Laird, younger son of the founder of the shipbuilding firm, and in 1832 promoted an expedition equipped with two armed steamships – one, in fact, an iron ship. Lander joined it, as one of its leaders, making a triumphant re-entry to some of the communities he had visited on his earlier, apparently more hazardous trip. He presented Chief Boy with the promised complement of muskets, together with various other presents, including full Highland dress. But the expedition faltered; malaria took its usual toll. There was hostility, especially from the people who controlled the Delta river traffic, who quite rightly foresaw serious competition. A ship's boat was fired on, and it is disconcerting to note Lander's comment: 'We found it necessary to chastise the natives and destroyed the town.' From neutral, even sympathetic, explorer, he had become capitalist exploiter. Lander took one ship a hundred miles up the Benue, and steamed up the Niger as far as Rabba; later, returning by canoe to join his ship, he was attacked by, according to his own calculations, '8000 or 10000 all armed with muskets and swords, Bonny, Benin and Brass people'. Escaping, wounded,

Lander stood up in his canoe, characteristically waved his hat, and gave 'a last cheer in sight of his adversaries'. He died of his wounds a week later in Fernando Po.

The lure of the Niger continued. The confluence of the Niger and Benue, first seen by Lander, became the site of an extraordinary experiment, the grand 'Philanthropic Expedition' of 1841. Here, the mythological vision of Africa as a potential paradise, a site where racial harmony and Christianity could be nurtured, if properly organised by the British, triumphed over any more commonsense appreciation of the difficulties. The enterprise was a bizarre example of misplaced Victorian enthusiasm, in which almost every facet and motive which propelled these expeditions is represented. There was engineering expertise and technology: John Laird constructed three paddle steamers, the *Albert*, the *Wilberforce* and the *Habib-es-Soudan*, the 'Friend of the Blacks', while Dr Reid designed a special ventilation system, vetted by Brunel. There was royal patronage and empire: Prince Albert was present at a valedictory meeting at Exeter Hall, inspected the ships at Woolwich, and presented their captains with gold chronometers; four queen's commissioners were appointed, and the expedition was voted funds by Parliament and placed under naval discipline. There was medicine, including the presence of Dr Thomson, whose self-experiments would prove that regular doses of quinine protected against malaria; there was the Church – including the young Samuel Crowther. There was science: Dr Vogel, the botanist, and four other geologists and naturalists; and there was a West Indian, Alfred Carr, who was to be in charge of the Model Farm. The Model Farm was the idea of the philanthropists, a way of helping to stem the effects of the slave trade by proving that a settled, industrious way of life could provide happiness and a modest income through profitable trade with the British. The expedition, which started with such bright hopes, foundered on malaria; and the *Wilberforce* rescued the remnants of the settlement in 1842. An official report was published in 1848, to be mercilessly mocked by Charles Dickens in *Bleak House*, where Mrs Jellyby's eyes are always fixed on Africa: 'We hope by this time next year to have from a hundred and fifty to two hundred healthy families cultivating coffee and educating the natives of Borrioboola-Gha, on the left bank of the Niger.'

Dickens's caustic satire injects a harsher note into the context of Victorian scientific exploration. Already, in the decade before *The Origin of Species*, two competing attitudes can be identified : a relatively objective and positive approach, which sought knowledge about the land and the people, and one which regarded Africa as irredeemably savage, a continent to be surveyed as a prelude to European domination.

Two expeditions which penetrated unexplored zones of Africa in 1850 illustrate this divergence and provide a sharp contrast in methods and personality. Francis Galton, cousin of the Darwins, with one of the acutest scientific minds of his time, set his mind on a serious trip in southern Africa. Galton had already visited Khartoum, where he went to call on an extraordinary Englishman: 'a white man nearly naked, as agile as a panther, with head shorn except for the Moslem tuft, reeking with butter, and with a leopard skin thrown over his shoulder'. The man turned out to be Mansfield Parkyns, fresh from his remarkable travels in Abyssinia, who had briefly been at Trinity College, Cambridge with Galton. Galton adopted Parkyns's custom of wearing local dress, and learned about camping and living rough. In April 1850 he set out with a young Swedish naturalist, Charles Andersson, and a fistful of introductions from the Royal Geographical Society. The Boer revolt blocked his intended route to Lake Ngami, so he sailed up the west coast to Walfisch Bay, planning to travel up the valley of the River Swakop to Damaraland. Galton proved an extremely tough and determined explorer, as he led his train of bullock wagons inland, eating the sheep and oxen as they went along. He had a brisk way with opposing lions, which were promptly shot, and with any local chief who offered resistance. The Namaquas were engaged in warfare with the Damaras, and had 'recently determined that no white man should pass through their country to the interior'. Galton rode to confront Jonker, the chief, mounted on an ox and wearing his hunting pink. 'I rated him soundly, in English first, to relieve my mind, and then in Dutch through my interpreters, brandishing my paper with the big seal, and thoroughly frightened him.' Jonker was apparently taken by surprise at this performance, and agreed to Galton's demands. The same imperiousness governed Galton's conduct of the expedition. 'I had to hold a little court of justice on most days, usually followed by corporal punishment, deftly administered. At a signal from me the culprit's legs were seized from behind, he was thrown forward on the ground and held, while Hans applied the awarded number of whip strokes.' This rough and ready justice 'became popular', adds Galton; he does not say with whom.

Galton had little interest in the flora and fauna of the country, and not much more in the people through whose lands he marched. The Damaras were for the most part 'thieving and murderous, dirty and of a low type'; their language, he claimed, contained not one word for gratitude, but fifteen 'that express different forms of villainous deceit'. The Ovampo whom he invaded next, he found more congenial, though he offended the chief Nangoro, when he presented Galton with a temporary wife: 'I found her installed in my tent in negro finery, raddled with red ochre and butter, and as capable of leaving a mark on anything she touched as a well-inked printer's roller. I was dressed in my one well-preserved suit of white linen, so I had her ejected with scant ceremony.' Supplies began to dwindle, and it seemed that they might be outstaying their welcome. When Galton applied for permission to leave, he was ordered to retrace his tracks, and they made their way back safely to Walfisch Bay. For his geographical achievements, in opening up and mapping an unknown area of central south Africa, Galton was awarded one of two annual gold medals by the Royal Geographical Society. Other honours followed, including a Fellowship of the Royal Society; and in 1878 Galton published a much used *Hints for Travellers* on the basis of his experience in the field. For him, though, Africa remained a land to be first explored, and then exploited. His innate sense of superiority would be expressed later in his theories about eugenics. His 1855 book *The Art of Travel* is full of hard-headed, practical hints – he advocated Mansfield Parkyns's technique for keeping clothes dry in the rain, which was to take them all off and then sit on them – and he was a fearless and ingenious survivor; but one senses that he was one of the few travellers who remained quite impervious to their surroundings, as narrowly English and imperial in his attitudes as he was when he set out.

Arguably the most impressive expedition, and the one with, in the end, the purest and most objective scientific purpose, was that of Heinrich Barth between 1850 and 1855. Barth's place on this official British venture was secured by the intervention of the Prussian ambassador to England, the Egyptologist Chevalier Christian von Bunsen, who was keen to promote a German scientific element. Barth, who had studied at the University of Berlin and obtained a lectureship there, had already undertaken one enterprising and decidedly dangerous journey in the Sahara, when he narrowly survived an attack by bedouin (one bullet, in his left thigh, remained there for the rest of his life). Barth's father, worried about the dangers, persuaded him to withdraw, which he reluctantly agreed to do. At this point Adolf Overweg, a geologist and astronomer, was attached to the expedition, but the British government insisted that Barth, having once volunteered, had a legal duty to go, and Barth's father was overruled. The expedition was initially under the leadership of James Richardson, who had already made a successful Saharan journey. Richardson's motives for this new central African expedition, a 'political and commercial expedition to some of the most important kingdoms of Central Africa', were twofold: 'one principal, if necessarily kept somewhat in the background – the abolition of the slave trade; one subsidiary, and yet important in itself – the promotion of commerce by way of the great desert'. To these were added the scientific goals, to be undertaken by Barth and Overweg. Richardson, according to the plans, was to retrace his route after reaching Lake Chad, leaving the Germans with a number of possibilities, including, astonishingly, taking a route east, either to the Nile, or all the way across central Africa towards Mombasa, in which case they were generously authorised to draw on the English consul in Zanzibar for an additional £200. (When it came to the point Barth decided on Timbuktu.)

Barth brought to the task of exploration a scientific detachment which sets him apart from many of his contemporaries. He was trained in geography and history; his skill in languages, beginning from a base of fluent Arabic, allied with his relentless curiosity and appetite for facts, makes him appear more like a twentieth-century anthropologist or ethnographer. Although he was attached to a British expedition, and was, effectively, its head for a significant period, he did not look at Africa and Africans from an imperial or insular point of view. He did not march into Kano in full officer's uniform, like Captain Clapperton, or lead a chorus of 'Rule Britannia' canoeing down the Niger like Lander. He wore Arab or 'Sudanic' costume; he travelled on camel – his faithful Bu-Sefi – or horseback, or on foot; he talked Hausa to his servants, learned the local languages, observed the local customs. He had with him a copy of the Koran, and, whenever it was awkward not to do so, would say his morning prayers like a Muslim. Partly because of the length of time he spent in one region, partly because of the nature of his intellectual training, partly because of his temperament, he understood to a far higher degree than his predecessors the structure of government in the kingdoms of Sokoto and Bornu, and the workings of the economy. He experienced frustrations and irritations, and periods of sickness, like any other traveller; he had his share of narrow escapes and bizarre encounters; but he did not allow these to dominate, or to deflect him from his main objectives. As soon as he had solved a difficulty, or recovered his health, he would be back at his task, observing, recording, measuring.

He also, one senses, liked to be on his own, which was just as well. The first leg of the journey across the desert had not been at all straightforward. For two days, Barth was lost; he had set off to climb a mountain ridge, become disorientated and exhausted, and was lucky to be found by one of the Berber servants. At a slightly later stage, their Tuareg escort threatened to kill all the Europeans if they did not immediately convert to Islam. But this did not deter Barth from making a detour to Agades on his own, and in January 1851, partly because of the uncertain state of their finances, the three decided to split up 'in order to try what each of us might be able to accomplish single-handed and without ostentation, till new supplies should arrive from home'. (New supplies could only reach them in response to a request which would be taken by caravan to Tripoli.) 'I now went on alone,' Barth commented, 'but felt not at all depressed by solitude, as I had been accustomed from my youth to wander about by myself among strange people.' He felt the benefits of civilisation, and enjoyed a splendid supper 'of a fowl or two, while a solitary maimol cheered me with a performance on his simple three-stringed instrument, which, however monotonous, was still expressive of much feeling, and accompanied with a song in my praise'.

Kano was Barth's objective – he describes it as 'this African London, Birmingham, and Manchester' – and he marks its importance in his narrative by a description of his modest little caravan as he approached his goal. It consisted of 'a very lean black horse, covered with coarse wool-like hair, worth four dollars, or perhaps less; a mare, scarcely worth more in its present condition; a camel, my faithful Bu-Sefi, evidently the most respectable four-footed member of the troop, carrying a very awkward load, representing my whole travelling household, with writing-table and bedding boards; a sumpter-ox, heavily laden; then the four human bipeds to match, viz. one half-barbarized European, one half-civilized Goberawi Tunisian mulatto, a young lean Tebu lad, and my stout, sturdy, and grave overseer from Tagalel'. They saw in the distance the top of the hill Dala, the real landmark of Kano: 'Kano had been sounding in my ears now for more than a year; it had been one of the great objects of our journey as the central point of commerce, as a great storehouse of information, and as the point whence more distant regions might be most successfully attempted. At length, after nearly a year's exertions, I had reached it.'

Barth formed friendly relations with the Governor of Kano, and spent his time in the city researching its history, analysing the trade and industry, and describing the structure of government and the taxation system of the province. He then set off for Bornu, which was to be the base for his next major enterprise, the locating of the River Benue. For part of the journey, he accompanied a 'most noble' Arab merchant from Morocco, enjoying the luxuries which his gentleman-like companion could afford: sitting in the shade of some tamarind trees, while a female attendant produced 'a variety of well-baked pastry, which he spread on a napkin before us, while another of the attendants was boiling the coffee'. Later in the journey he learned that James Richardson had died, and visited (Barth's inverted commas) his 'white man's grave'. Barth satisfied himself that Richardson had had a decent burial, and gave a small present to a man who promised to take care of the grave, but he seemed rather more upset at having to part with his favourite camel, to spare it the effects of the rainy season. In Kukawa he entered into negotiations with the sheikh, and especially with his vizier el Haj Beshir ben Ahmed Tirab. His long-term objective was to draw up a treaty on behalf of the British government, but there were many lengthy preliminaries, including taking into his own control all Richardson's possessions which, understandably, had been appropriated.

Barth left a subtle and sympathetic portrait of Mohammed el Beshir. He was the son of the most influential man in Bornu after the sheikh, and he 'enjoyed all the advantages which such a position could offer for the cultivation of his mind, which was by nature of a superior cast. He had gone on a pilgrimage to Mekka, by way of Ben-Ghazi, when he had an opportunity both of showing the Arabs near the coast that the inhabitants of the interior of the continent are superior to the beasts, and of getting a glimpse of a higher state of civilization than he had been able to observe in his own country.' Barth judged him 'a most excellent, kind, liberal, and just man', who 'might have done much good to the country, if he had been less selfish and more active'. Politically, he was neglectful, failing to ingratiate himself with the more powerful courtiers and indulging himself in his interests, including an unusual harem – a 'credulous person might suppose that he regarded his harim only from a scientific point of view; – as a sort of ethnological museum . . . which he had brought together to impress upon his memory the distinguishing features of each tribe'. Muhammed el Beshir's intellectual curiosity matched Barth's, and it was through him that Barth was allowed to see the historical records of Edris Alawoma – though he would not allow Barth to do more than read them over his shoulder. This intimacy might have been fatal to Barth; Beshir was killed in Barth's absence, as part of a coup against the ruler, but by the time Barth returned, the old sheikh had been restored to power.

Barth was joined in May by a fatigued and distinctly unwell Overweg. Barth does not sound overjoyed to see him: Overweg had only the clothes he was wearing: 'I was therefore obliged to lend him my own things, and he took up his quarters in another part of our house, though it was rather small for our joint establishment.' With Richardson dead, Barth and Overweg took up the negotiation of a treaty with the sheikh and his vizier: 'Both of them assured us of their ardent desire to open commercial intercourse with the English, but at the same time they did not conceal that their principal object in so doing was to obtain fire-arms. They also expressed their desire that two of their people might return with us to England, in order to see the country and its industry, which we told them we were convinced would be most agreeable to the British Government.' The local Arab traders, worried, in Barth's view, by the long-term threat to their trade, immediately spread a rumour that seven English vessels had arrived at Nupe, on the Niger, a story which was soon proved to be false, but which cooled the friendly atmosphere a little. The treaty was finally signed by Barth on 3 September 1852, though little came of it, as British policy had shifted by the time Barth returned to England, and the government reneged on their overtures to Bornu, Sokoto and Timbuktu.

Next Barth turned his attention to one of the great objectives of his journey: the course of the River Benue. For once, the reality matched his imagination, for he came on the river where it was joined by the Faro. His accurate, scientific description is enriched by a sense of triumph, and an optimistic vision of the future:

> The principal river, the Benuwe, flowed here from east to west, in a broad and majestic course, through an entirely open country, from which only here and there detached mountains started forth. The banks on our side rose to twenty-five, and in some places to thirty feet, while just opposite to my station, behind a pointed headland of sand, the Faro rushed forth, appearing from this point not much inferior to the principal river, and coming in a fine sweep from the south-east, where it disappeared in the plain, but was traced by me, in thought, upwards to the steep eastern foot of the Alantika. . . . I looked long and silently upon the stream; it was one of the happiest moments of my life.
>
> I had now with my own eyes clearly established the direction and nature of this mighty river; and to an unprejudiced mind there could no longer be any doubt that this river joins the majestic watercourse explored by [Park and Lander]. Hence I cherish the well-founded conviction, that along this natural highroad European influence and commerce will penetrate into

> the very heart of the continent, and abolish slavery, or rather those infamous slave-hunts and religious wars, destroying the natural germs of human happiness, which are spontaneously developed in the simple life of the pagans, and spreading devastation and desolation all around.

Barth gave himself the rare luxury of a river bath – he had not yet reached the wise conclusion that river-bathing in tropical Africa was not good for a European (or anyone else) – before crossing the river precariously by canoe, and continuing his journey to the town of Yola, the capital of Adamawa and the furthest point south of his travels.

Barth's vision as he gazed at the Benue was essentially in tune with the philanthropic dream of 1841. The site of the confluence of the Niger and the Benue first identified by Lander continued to attract as a potential centre for commerce and Christianity, the two elements which would be linked inextricably for the British by the example and philosophy of David Livingstone. Livingstone, who had discovered Lake Ngami in 1849, would direct his later searches towards the sources of the Nile. But his vision for Africa, as expressed in his address at Cambridge in 1857, was one which pleaded to the West to keep Africa 'open' : 'Do not let it be shut again! I go back to Africa to try to make an open path for commerce and Christianity; do you carry out the work which I have begun.'

This phase of the exploration of the Niger concludes with the 1854 expedition led by Dr Baikie, a combined venture once again under Admiralty orders but with an important contribution from Macgregor Laird. Its primary purpose was to explore further up the Benue, beyond the point reached by Lander in 1832; and also to try and locate Dr Barth, who, with Vogel, was supposedly somewhere in the Lake Chad region, but out of communication. Among the twelve Europeans was a zoologist, John Dalton, and the long-term objective was to introduce 'legitimate' trade and the Christian religion. Learning from Dr Thomson's experiments, 'quinine wine' was taken regularly to prevent malaria. On board was Samuel Adjai Crowther, by this time ordained as the first modern African minister. The expedition spent sixteen weeks on the Niger and Benue, achieved its geographical objectives, did enough trading to encourage future prospects, and was 'remarkable for the safe return to this country of all Europeans engaged in it'. Baikie, like Lander before him, was enthralled by his vision of Africa. He returned to the Niger in 1857, and stayed on to found a miniature colony in Lokoja, while on the other side of the river

Samuel Crowther, the first bishop of the Niger, spread the Gospel from his base at Igbobi. Baikie, marooned on an island in Nupe in October 1857 awaiting a relief ship, did his scientific duty as one of Darwin's far-flung correspondents:

> During our voyage up the river, & our residence in this spot I have been paying considerable attention to domestic breeds of animals, & I hope to be able to bring home specimens of most of them, & full accounts of all. I am in great hopes of being able to make a run as far as Princes' or St Thomas' before I return, as a Zoological & Botanical visit to these islands would be most interesting. I do not see why these islands should not contain some indigenous & peculiar species.

In spite of the fervent attachment of men like Baikie and their attempts to live permanently in central Africa, the consensus, especially after the failure of the settlement, remained that only Africans could take root there. The overarching plan was for Africans who had been uprooted by the slave trade, and later freed and educated – and, from a European standpoint, converted – to return and, with British help, revitalise the country which had been devastated by the long-term impact of the slave trade. Yet Baikie stayed on, committing himself and his health to the cause, living on threepence a day, and eventually adopting a different pattern of life, with wives and children. The arguments he put to the Foreign Office for continued support were the arguments of interest and empire: 'What I look to are the securing for England a commanding position in Central Africa'; 'Were a beginning once made, there is nothing which English enterprise and capital may not effect as they have so often done in much more unpromising and unfruitful regions.' In an earlier plea, he had argued how essential a properly armed vessel was: 'Mr Laird was perfectly correct when he defined "moral force in Africa" as meaning a thirty-two pounder with an English sailor standing behind it.' (This show of force was to sort out the Delta people.) Baikie, absorbed in constructing a Fulani grammar, lost his fluency in English; John Whitford, a trader who visited the settlement in the 1860s, referred to Baikie's successors as 'the lonely ones', and to their houses as 'Her Majesty's Mud-Huts'. Whitford, too, could be entranced by the European myth of Africa: 'After sunset, when the air becomes pleasantly cool, hilarity, song, and rollicking fun come forth from the conical huts. Young men and maidens assemble, and, with their arms as Nature formed them, twined around each other's necks, they softly breathe the old, old story which Adam and Eve first told in the Garden of Eden.' Music begins, and dancing, in which everyone joins until midnight, 'when all retire to their respective huts and enjoy a grand gorge of good food, washed down by calabashes of palm-wine, and after that they fall asleep. Verily, the natives of luxuriant Central Africa are very happy and contented in their natural state.' However, Whitford, a chapter later, is proposing the constructing of a railway from Lagos to Lokoja. The Romans, by building roads, did a similar kindness 'for the benefit of our tattooed and painted, wild-skin clothed and ragged, but sturdy ancestors. We ought to extend the same kindness to African savages . . . '. When Barth returned successfully to London and reported on the possibility of a treaty with the well-established nations of the interior, he was kindly received by Lord Palmerston, but his suggestions met with little favour. 'I have the satisfaction to feel,' he concluded his three volumes of immensely detailed and well-documented travels, 'that I have opened to the view of the scientific public of Europe a most extensive tract of the secluded African world.' 'The scientific public of Europe' is a phrase which matches Barth's objectivity and his passion for knowledge. His hopes were not realised. The Niger became first a trading route, a way of bypassing the coastal middlemen; and then an imperial highway, like the Congo and the Nile. Josiah Wedgwood's abolitionist cameo, with the African suppliant exclaiming 'Am I not a Man and a Brother?', was easily forgotten, or given a qualified answer, 'Not yet.' The evidence was complex. Even with the advent of quinine, the struggle to remain healthy, let alone overcome other difficulties, confused the issues. Visions of Eden, of the abundance and fecundity of nature, of the sophisticated economies and societies within Africa, were relegated. It was much more convenient to ignore all signs of civilisation, and instead to promote the idea of the savage; to adopt the attitude of Galton, rather than Barth.

3 The Naturalists in the Amazons

IN CAPTAIN HISLOP'S HOUSE IN Santarém one evening in 1849, eight hundred miles up the Amazon river, three scientific explorers dined together: Henry Walter Bates, Alfred Wallace and Richard Spruce. The Amazon was the site of the greatest concentration of British naturalists in the middle of the century. Their long-term inspiration, as with so many naturalists and explorers, came from Humboldt. The immediate impetus came from an 1847 publication by an American, W.H. Edwards, *A Voyage up the River Amazon*, which emphasised the ease of access and the healthy climate. Edwards – who described his forty-day voyage as a 'jolly cruise', in a 'sort of pleasure craft' stocked with pots of New York oysters, Yankee doughnuts and tin cases of cheese – had an eye for a journalistic phrase: the mightiest of rivers rolled majestically through primeval forests of boundless extent, in a land of Amazonian women, cannibal Indians and epicurean anacondas. He also provided a useful checklist for the aspiring traveller, and was full of praise for the Amazon region: the climate was healthy, the forest luxuriant, the people friendly, the cost of living cheap. Besides, it was comparatively close – a month's sail, in good conditions. More importantly, South America had not yet been widely collected: for so long under Spanish and Portuguese influence, it was only since Humboldt and Bonpland that it had been opened up to northern European travellers. Among English naturalists there was the eccentric Charles Waterton, 'seized with an unconquerable aversion to Piccadilly', whose *Wanderings in South America* was greeted in 1825 with a mixture of delight, disbelief and derision. Written in a style which might have been learned from Laurence Sterne, Waterton reported his hands-on encounters with snakes, caymans and sloths, and his investigations into the properties of curare and other toxics; overwhelmingly, he conveyed his enthusiasm for the natural world, and his complete fearlessness and confidence in living with the Indians of the Orinoco region. The more systematic naturalist Karl von Martius, the Professor of Botany at Munich, had explored Brazil with von Spix from 1819 and written pioneering books about South American

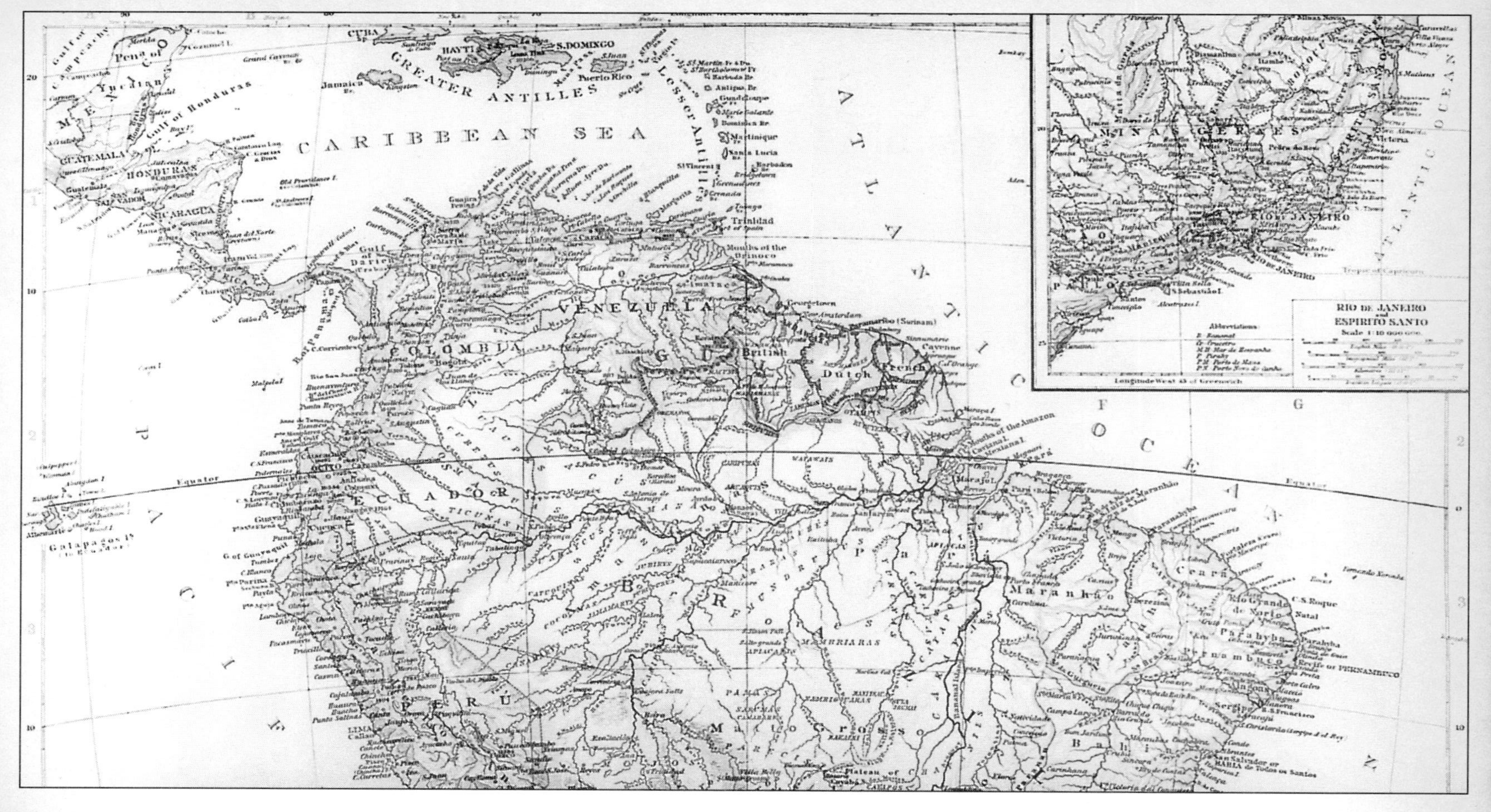

The Amazon region (The Times Atlas, 1895)

plants; but all these simply indicated the apparently inexhaustible botanical riches which were waiting to be discovered, a statement which is almost as true today as it was 150 years ago. The Amazon was the naturalists' El Dorado.

Alfred Wallace would, ten years later, formulate quite independently of Darwin the idea of natural selection as the key to evolution. Henry Bates, the entomologist whose speciality was mimicry among insects, became the secretary of the Royal Geographical Society. Richard Spruce would eventually retire with a pension for his contribution to procuring cinchona seeds and plants, broken in health though not in mind or spirit: his fieldwork still provides stimulus and inspiration for botanists and anthropologists.

Wallace was, by his origins, an apparently unlikely candidate to make any major scientific breakthrough. He grew up against a background of financial insecurity; his father, restless and feckless, moved from place to place and occupation to occupation with little purpose, and Wallace had to leave Hertford grammar school in 1836 at the age of fourteen, when funds ran out. He was sent off to the care of elder brothers, first to London to work alongside John, an apprentice carpenter. There he attended lectures at the Working Men's Institute, and came across the ideas of Malthus and of Robert Owen, each crucial to him in different ways. From Owen he first absorbed the utopian socialist vision which remained with him all his life, influencing his thinking, and shaping the way he looked at the world and interacted with the societies he encountered. Owen's thinking sprang equally from the Enlightenment and from the harsh realities of the Industrial Revolution. For Owen, the industrial structures of society could be transformed to obtain freedom, rather than enslavement, for those who worked in them; each individual was the product of circumstance, and circumstances could be changed to produce an earthly paradise. He preached the essential unity of mankind, believing that the age offered to the family of man 'without a single exception, the means of endless progressive improvement, physical, intellectual, and moral'. Owen emphasised the fundamental importance of education, of knowledge, to effect moral change, and managed to make his model commune at New Lanark work financially. This brief but vivid exposure to an alternative view of the world was something that Wallace would never have come across so early if he had followed, say, Darwin's smooth progression from Shrewsbury School to a Cambridge college. Owen's was both a doctrine and an attitude that permeated Wallace's future thinking. At the same time he was introduced to Malthus's *Essay on the Principle of Population* (1798). The Revd Thomas

Malthus, an orthodox clergyman and economist, could not accept that men might, of their own efforts, succeed in improving or perfecting themselves; such attempts could only serve to disturb the ordained balance of nature, in which built-in checks such as disease, famine and war keep down the growth of population to what the food supply can sustain; without these checks, no society would be able to survive. The spirit of Owen, and the brutal mechanism of Malthus, remained in Wallace's mind.

Wallace's third phase of education stemmed, like the Working Men's Institutes, from the industrialisation of Britain. He was transferred to the care of another brother, William, fourteen years older, who was a land surveyor. The relentless mapping of the rest of the world mirrored on a larger scale what had been happening within Britain and Ireland. The canal system and, later, the railways were two of the areas which provided jobs for a new profession. Wallace learned to use the tools of the trade: the sextant, the compass, the theodolite. He learned at first hand about fossils, and he had time and opportunity to extend his interests in natural history as he moved around the country with his brother, from village to village. With no other distractions, he spent his Sundays in solitary walks, looking, collecting and, once he had acquired a guide to plants, classifying. He taught himself botany. When he was twenty-one he left surveying for a while and took a job teaching at the Collegiate School in Leicester; there, through a chance – though wholly logical – encounter at the public library, he met Henry Bates, who astonished his fellow enthusiast by his collection of hundreds of beetles. The two men became friends, and went on beetle-hunting trips together. When Wallace returned for a time to surveying, in the great railway boom of the 1840s, they continued to correspond, and to visit each other; they exchanged notes on their reading – Robert Chambers's *Vestiges of the Natural History of Creation*, Darwin's *Journal*, Lyell's *Principles of Geology*. In 1847, Wallace read Edwards's *A Voyage up the River Amazon*. That autumn, Bates recalled, Wallace proposed 'a joint expedition up the river Amazons, for the purpose of exploring the natural history of its banks; the plan being to make for ourselves a collection of objects, dispose of the duplicates in London to pay expenses, and gather facts, as Mr Wallace expressed it in one of his letters, "towards solving the problem of the origin of species", a subject on which we had conversed and corresponded much together'.

Bates, Wallace and Spruce belonged to a new breed of scientist. They were not sponsored directly by the government, like Huxley or Darwin, attached to Royal Naval survey ships; they were not salaried, like the plant-hunters employed by the nurserymen, or even promised a reward on their safe return, like Richard Lander. They were scientific entrepreneurs, trading in beetles and birds and monkeys and dried plants, who needed to collect extensively even to pay their expenses, let alone secure a possible income for the future, when they might hope to work up their private collections and live off a store of knowledge and fieldwork rich enough to last the rest of their lives. The British Museum assured them there would be a good market for their collections. They made arrangements with a London agent and dealer, Samuel Stevens, who had premises in Bedford Street, just round the corner from the British Museum – an excellent choice, as it turned out. Stevens became Wallace's lifeline, not just for this trip, but for his later and longer journeys in the Malay Archipelago.

> He continued to act as my agent during my whole residence abroad, sparing no pains to dispose of my duplicates to the best advantage, taking charge of my private collections, insuring each collection as its despatch was advised, keeping me supplied with cash, and with such stores as I required, and, above all, writing me fully as to the progress of the sale of each collection, what striking novelties it contained, and giving me general information on the progress of other collectors and on matters of general scientific interest.

In addition, Stevens arranged for the publication of Wallace's letters in the *Annals and Magazine of Natural History*, and attended meetings of the Entomological Society at which he advertised Wallace's and Bates's latest finds. Even at this early stage, Wallace and Bates were moving out of the journeyman role of collector towards that of scientific thinker and analyst.

Over the succeeding years, Wallace, Bates and Spruce made the Amazonian rain forest their home, exploring and collecting, observing and recording with amazing fortitude and endurance but also, in spite of the inevitable privations and bouts of ill-health, with an infectious sense of pleasure and wonder.

On 20 April 1848 Wallace and Bates left Liverpool on the *Mischief*, which at the end of May came to anchor opposite the southern entrance to the River Amazon. The pilot went on board, and they sailed upriver. 'Early on the morning of the 28th,' wrote Wallace, 'we again anchored; and when the sun rose in a cloudless sky, the city of Pará (Belém), surrounded by the dense forest, and overtopped by palms and plantains, greeted our sight, appearing doubly beautiful from the presence of those luxuriant tropical productions in a state of nature, which we had so often admired in the conservatories of Kew and Chatsworth.' Bates wrote:

> It was with deep interest that my companion and myself, both now about to see and examine the beauties of a tropical country for the first time, gazed on the land where I, at least, eventually spent eleven of the best years of my life. To the eastward the country was not remarkable in appearance, being slightly undulating, with bare sand-hills and scattered trees; but to the westward, stretching towards the mouth of the river, we could see through the captain's glass a long line of forest, rising apparently out of the water; a densely-packed mass of tall trees, broken into groups, and finally into single trees, as it dwindled away into the distance. This was the frontier, in this direction, of the great primaeval forest characteristic of this region, which contains so many wonders in its recesses, and clothes the whole surface of the country for two thousand miles from this point to the foot of the Andes.

Bates and Wallace spent their first months working together. They made Pará their base, and from there they trawled the forest pathways, or, as they came to know more people, accepted invitations to visit plantations and settlements further afield by boat. Bates's enthusiasm is immediately apparent: the profusion of life, and its extraordinary diversity, struck him, beginning with 'the mingled squalor, luxuriance and beauty' of the women, an example of the striking mixture of natural riches and human poverty (by 'squalor', he is referring to their 'slovenly' dress, and the fact that many went barefoot even in the 'city'). 'But amidst all,' he goes on, 'and compensating every defect, rose the overpowering beauty of the vegetation. The massive dark crowns of shady mangoes were seen everywhere amongst the dwellings, amidst fragrant blossoming orange, lemon, and many other tropical fruit trees; some in flower, others in fruit, at varying stages of ripeness. Here and there, shooting above the more dome-like and sombre trees, were the smooth columnar stems of palms, bearing aloft their magnificent crowns of finely-cut fronds.' All this was wonderful to the newly arrived travellers, whose last nature ramble had been over the bleak Derbyshire moors on a sleety morning in April.

Bates's enjoyment, and his broadly sympathetic response to the people he moved among, is a striking feature of his book. He was soon at work, analysing the behaviour of his specialist collecting group, the insects: there were 700 species of butterfly to be found within an hour's walk of Pará, numerous species of ant, not many ground but plenty of arboreal beetles, and many species of Mygale, 'monstrous hairy spiders' – 'One very robust fellow, the Mygale Blondii, burrows into the earth, forming a broad slanting gallery, about two feet long, the sides of which he lines beautifully with silk. He is nocturnal in his habits. Just before sunset he may be seen keeping watch within the mouth of his tunnel, disappearing suddenly when he hears a heavy foot-tread near his hiding-place.' From his first days in the Amazons, it is clear that Bates was an acute observer of behaviour, a scientific thinker just as much as a collector. He was also, because of his need to peer closely in search of his favourite insects, brought into close quarters with every kind of living thing. Quite early on, he became entangled with a six-foot snake as he passed through a thicket, and trod on another highly poisonous specimen which fortunately bit his trousers rather than his leg: as he commented, it was rather alarming 'in entomologising about the trunks of trees, to suddenly encounter, on turning round, a pair of glittering eyes and a forked tongue within a few inches of one's head'. There was not much chance of rapid manoeuvre, with a double-barrelled gun over the left shoulder, net in the right hand, insect box suspended on the left, game bag on the right with thongs to hang lizards, snakes, frogs and birds, and a pincushion pinned to the shirt, with six sizes of pins.

The two men settled down in a country villa a mile or two from the city, with forest on three sides, and near a house where two of their famous predecessors, Spix and Martius, had stayed in 1819. They rose at dawn, and spent the two hours before breakfast in ornithology. Then from 10 a.m. to 2 or 3 p.m., it was entomology, 'the best time for insects in the forest being a little before the greatest heat of the day, 92 or 93 Fahrenheit'. By this time, 'every voice of bird or mammal was hushed', and their Indian and mulatto neighbours were asleep in their hammocks or seated on mats in the shade, as the collectors returned home fatigued from their ramble. They dined at four, took tea at seven, and spent most evenings preserving their collections, and making notes. This was the kind of relentless routine they were able to maintain for month after month and, in Bates's case, for over eleven years. It was a slow way of making a living, at fourpence a specimen less 20 per cent commission and 5 per cent for transport and insurance. Bates's profit for one twenty-month period was just short of £27.

The two stayed in close touch for most of their first year, and undertook one joint expedition up the River Tocantins together. Then they decided to separate, apparently amicably, Wallace to explore the Rivers Guamà and Capim, and Bates to return to Cametá on the Tocantins. He lived for a time with an Indian family, and went hunting with another Indian friend, Raimundo, whom he got to know well. Raimundo's 'only complaint' against the white man 'was that he monopolised the land without having any intention or prospect of cultivating it. He had been turned out of one place where he had squatted and cleared a large piece of forest. . . . Raimundo spoke of his race as the red-skins, "pelle vermelho"; they meant well to the whites, and only begged to be let alone. "God," he said, "had given room enough for us all."' The party arrived at the hunting ground about half-past four, and slept on the benches of the canoe while the boat drifted with the tide.

> My clothes were quite wet with the dew. The birds were astir, the cicadas had begun their music, and the *Urania Leilus*, a strange and beautiful tailed and gilded moth, whose habits are those of a butterfly, commenced to fly in flocks over the tree-tops. Raimundo exclaimed, "Clareia o dia!" – "The day brightens!" The change was rapid: the sky in the east assumed suddenly the loveliest azure colour, across which streaks of thin white clouds were painted.

The expedition then became entirely practical, as they hunted Pacas and Cutías for meat – Bates was finding he could not live comfortably on vegetable food alone, and disliked the salt-fish which was the normal supplement. Special moments like the one above permeate his narrative. Bates seems to have had an unusual empathy for the people as well as for the forest, and a sense of the rhythm of life to which he is able to attune himself. From time to time, he sees a glimpse of an ideal way of existence which can be attained in this special environment:

> The harsh, slave-driving practices of the Portuguese and their descendants have been the greatest curses to the Indians; the Mundurucús of the Cuparí, however, have been now for many years protected against ill-treatment. This is one of the good services rendered by the missionaries, who take care that the Brazilian law in favour of the aborigines shall be respected by the brutal and unprincipled traders who go amongst them. I think no Indians could be in a happier position than these simple, peaceful, and friendly people on the banks of the Cuparí. The members of each family live together, and seem to be much attached to each other; and the authority of the chief is exercised in the mildest manner. Perpetual summer reigns around them; the land is of the highest fertility, and a moderate amount of light work produces them all the necessaries of their simple life.

He was also quite calm about the occasional dangerous encounter:

the anaconda who attacked his canoe at night, forcing its way into the chicken coop and making off with a couple of hens, turned out, when it was killed a few days later, to be a not very large specimen, only eighteen feet nine inches long. Soon after, he pursued a huge boa constrictor whose body resembled 'a stream of brown liquid' to take a note of its length and markings, only to see it disappear into a dense swampy thicket, where even Bates chose not to follow.

Bates's next base was at Obydos, fifty miles or so further upriver from Santarém, where Wallace was temporarily installed; from there, he travelled to Barra (present-day Manaos), which was sited close to the junction of the Rio Negro and the Solimões, or upper Amazon. There he enjoyed a few weeks' holiday, and a few weeks of Wallace's company, rambling through the forest, or visiting the local beauty spots:

> The waters of one of the larger rivulets which traverse the gloomy wilderness here fall over a ledge of rock about ten feet high. It is not the cascade itself, but the noiseless solitude, and the marvellous diversity and richness of trees, foliage, and flowers, encircling the water basin, that form the attraction of the place. Families make pic-nic excursions to this spot; and the gentlemen – it is said the ladies also – spend the sultry hours of midday bathing in the cold and bracing waters. The place is classic ground to the naturalist, from having been a favourite spot with the celebrated travellers Spix and Martius during their stay at Barra in 1820.

Classic or not, there were not enough birds or insects to hold Bates's interest long. Soon he and Wallace were planning the next stages of their collecting trip, splitting up the country between them – Wallace to the Rio Negro, and the ascent of its tributary the Uaupés, Bates to the next town of any importance on the Solimões, Ega, some 400 miles, which 'we accomplished in a small cuberta, manned by ten stout Cucáma Indians, in thirty-five days'. He made Ega his headquarters for the next four and a half years, making excursions from it as far afield as 400 miles, and once going all the way downstream to Pará to supervise a shipment of his collections back to England – on this visit he contracted yellow fever, but managed to dose himself through it. (He failed to save the life of Herbert Wallace, Alfred's younger brother, who was preparing to go back to England.) The Amazon, or some parts of it, was proving less healthy than the naturalists had imagined.

In Ega, Bates followed his pursuit 'in the same peaceful, regular way as a Naturalist might do in a European village'. He had a workshop and study in a dry, spacious cottage, with a large table and his small library of reference books on shelves in rough wooden boxes; cages for drying specimens were hung from the rafters by cords smeared with a bitter vegetable oil to deter the ants, and rats and mice were foiled by inverted bowls strategically placed halfway down. He rose with the sun, and walked down grassy streets wet with dew to bathe in the river, keeping one eye fixed on the alligators who in the dry season tended to lurk in wait for 'anything that might turn up at the edge of the water: dog, sheep, pig, child, or drunken Indian'. Then five or six hours every morning were spent in collecting in the forest, and the hot hours of the afternoon, or the whole day if it was rainy, were occupied in preparing and ticketing the specimens, making notes, dissecting, and drawing. There were few dangers: apart from the alligator precaution, the only animals to be much feared were the snakes – common enough in the forest, but, as Bates remarked, 'no fatal accident happened during the whole time of my residence'. There was no danger from the inhabitants 'in a country where even incivility to an unoffending stranger is a rarity'. The main drawbacks were ill-health from bad and inadequate food, and the irregularity of news from 'the civilised world down river'. It was, in fact, the lack of intellectual society and the 'varied excitement of European life' which he felt most acutely, an absence which slowly increased in its impact during his years there until it became almost insupportable. Sometimes he had a supply by steamer every three months or so; then he would parcel out his stock of reading with strict economy, going through each number of the *Athenaeum* three times, and ending up by reading all the advertisements from beginning to end. Correspondence was slow and uncertain; writing to Wallace, for example, by then somewhere in the Celebes, an act of faith. Once he received nothing for a whole year, at the end of which his clothes were in rags, his shoes had fallen to pieces and, with the additional misfortune of a robbery, his supply of copper money had run out. Having to go barefoot was a great inconvenience in tropical forests, 'notwithstanding statements to the contrary that have been published by travellers'. Bates was probably referring to the eccentric Charles Waterton, who claimed that he seldom wore shoes and stockings: 'In dry weather they would have irritated the feet, and retarded me in the chase of wild beasts; and in the rainy season they would have kept me in a perpetual state of damp and moisture.' But Waterton had a high pain threshold, claiming to have slept night after night with his foot trailing from his hammock in the hope that a vampire bat would come along and suck his blood.

On his trips upriver, Bates changed station from south to north bank, and noted that a considerable number of the species, especially of insects, were 'representative forms' – 'species or races which take the place of other allied species or races' – of others found on the opposite bank. He concluded that 'there could have been no land connection between the two shores during, at least, the recent geological period. . . . All these strongly modified local races of insects confined to one side of the Solimões (like the Uakaris [monkeys]) are such as have not been able to cross a wide treeless space such as a river.' Bates is writing in 1856, before *The Origin of Species*, and grappling with the same problems of variety and distribution which were engaging Wallace, by this point travelling in the Malay Archipelago, and which had emerged for Darwin after the *Beagle* voyage as one of the key questions posed by the fauna of the Galapagos Islands.

Bates, understandably after eleven years in the Amazons, was beginning to run out of energy. Intellectual isolation, erratic diet and ill-health had taken their toll, though he had been relatively fortunate in the last respect. The discomforts and hazards of the area begin to loom larger in his narrative: the mosquitoes and *piums* swarmed more ferociously, mould attacked his collections. Bates retained a keen sense of the ridiculous, though he made light of the hazards. One episode, whose illustration provided the frontispiece for his book, shows a bespectacled naturalist 'mobbed by curl-crested toucans'. Bates had shot one – it was wounded, and set up a loud scream when he attempted to seize it. 'In an instant, as if by magic, the shady nook seemed alive with these birds. . . . They descended towards me, hopping from bough to bough, some of them swinging on the loops and cables of woody lianas, and all croaking and fluttering their wings like so many furies.' However, this was not a case of the naturalist hunted by his prey; once the wounded bird was dispatched and the screaming stopped, they melted back into the trees and had the sense to disappear before Bates could reload and turn them into specimens.

A rare but severe attack of fever interrupted the relentless collecting. Bates sounds outraged that he should succumb – the ague, he insisted, did not exist in this favoured territory bordering the shores of the Solimões. He had to have recourse to a small phial of quinine, bought eight years before in Pará, which he mixed with warm camomile tea. 'The first few days after my first attack I could not stir, and was delirious during the paroxysms of fever; but the worst being over, I made an effort to rouse myself, knowing that incurable disorders of the liver and spleen follow ague in this country if the feeling of lassitude is too much indulged. So

8. Bates Mobbed by Curl Crested Toucans (or, Toucans attacked by Naturalist). (Frontispiece, *The Naturalist on the River Amazons*)

every morning I shouldered my gun or insect-net, and went my usual walk in the forest. The fit of shivering very often seized me before I got home, and I then used to stand still and brave it out.' This final bout shattered even Bates's resolve, and he abandoned his plans to shift 600 miles west to the Peruvian towns at the foot of the Andes, and instead retraced his route to Pará, bound for New York and England.

Learning from Wallace's bitter experience of a disastrous shipboard fire, he parcelled his collection into three and dispatched it by three separate ships to reduce the risk. On the evening of 3 June 1859 he 'took a last view of the glorious forest' for which he had so much love, and to which he had devoted so many years. As he left the Naturalist's Paradise, a crowd of unusual thoughts occupied him, recollections of the English climate and the English industrial landscape:

> Pictures of startling clearness rose up of the gloomy winters, the long grey twilights, murky atmosphere, elongated shadows, chilly springs, and sloppy summers; of factory chimneys and crowds of grimy operatives, rung to work in early morning by factory bells; of union workhouses, confined rooms, artificial cares, and slavish conventionalities. To live again amidst these dull scenes I was quitting a country of perpetual summer, where my life had been spent like that of three-fourths of the people in gipsy fashion, on the endless streams or in the boundless forests.

It was a dismaying prospect. Three years later, as he completed his book, *The Naturalist on the River Amazons*, for John Murray, Bates had adjusted his point of view, and could comment that he found civilised life 'incomparably superior' to the spiritual sterility of half-savage existence, even though passed in the Garden of Eden. However, it was only in its social aspect that the bleak North was superior to tropical regions: 'for I hold to the opinion that, although humanity can reach an advanced state of culture only by battling with the inclemencies of nature in high latitudes, it is under the equator alone that the perfect race of the future will attain to complete fruition of man's beautiful heritage, the earth'. Back in smoky, industrialised England, with a people striving for survival and improvement under crepuscular skies, Bates retained his memory of the well-balanced forces of nature, in this vision of a perfected race living in the equatorial forests of the Amazon.

Wallace's account of his briefer time in the Amazon was published in 1853, long before Bates's book. It was based only on the journals of the beginning and end of his trip: the notes of the central part, together with the bulk of his collections and sketches, were lost at sea. Again, as with Bates, Wallace's *Travels on the Amazon and Rio Negro* reveals the mind of a scientific enquirer, interspersed with the enthusiasm of the collector – 1,300 species of insects in two months – and the same sense of buoyant wonder at the luxuriance of nature. The appendage to the title, 'With an Account of the Native Tribes', acknowledges his strong anthropological interest; and the additional description, 'And Observations on the Climate, Geology, and Natural History of the Amazon Valley', indicates his claim to be an all-round scientist, not just a collector. As early as October 1848, he wrote:

> In all works on Natural History, we constantly find details of the marvellous adaptation of animals to their food, their habits, and the localities in which they are found. But naturalists are now beginning to look beyond this, and to see that there must be some other principle regulating the infinitely varied forms of animal life. It must strike every one, that the numbers of birds and insects of different groups, having scarcely any resemblance to each other, which yet feed on the same food and inhabit the same localities, cannot have been so differently constructed and adorned for that purpose alone.

The interaction between varieties, localities, food supplies, and the need for some underlying principle as regulator, occupied Wallace's mind as he collected, labelled and drew his finds. The word 'adorned' may imply that he was thinking about sexual selection. An alligator hunt in a shallow lake prompted more thoughts about populations:

> In fact, the abundance of every kind of animal life crowded into a small space was here very striking, compared with the sparing manner in which it is scattered in the virgin forests. It seems to force us to the conclusion, that the luxuriance of tropical vegetation is not favourable to the production and support of animal life. The plains are always more thickly peopled than the forest; and a temperate zone, as has been pointed out by Mr Darwin, seems better adapted to the support of large land-animals than the tropics.

In June 1849, in a passage on slavery – he was staying with an estate-owner whose slaves were 'as happy as children' – he introduced the phrase which would form part of his argument about evolution:

> Can it be right to keep a number of our fellow-creatures in a state of adult infancy, – of unthinking childhood? It is the responsibility and self-dependence of manhood that calls forth the highest powers and energies of our race. It is the struggle for existence, the 'battle of life', which exercises the moral faculties and calls forth the latent sparks of genius. The hope of gain, the love of power, the desire of fame and approbation, excite to noble deeds, and call into action all those faculties which are the distinctive attributes of man.

It is at first slightly disconcerting to find Wallace, the most humane of men, arguing not from moral grounds, but from a biological concept about the way the human race has developed. For him, the biological imperative took precedence: the moral sense followed as a natural consequence. To keep a slave was like keeping an animal in captivity: without liberty, the slave cannot progress beyond childhood, the animal part of man's existence.

By November 1849 Wallace, Bates and Spruce (and Wallace's younger brother Herbert) were all in the area around Santarém, 1,000 miles inland from Pará, enjoying the hospitality of the old Scotsman, Captain Hislop. From Santarém, they moved on to Barra. Wallace spent six slightly frustrating months there, waiting for the rains to end, waiting for a canoe he had ordered, waiting for letters. Finally, on the last day of August 1850, he said goodbye to his brother: Herbert, clearly not cut out for the life of a collector, was to stay in Barra for another six months before returning to England. Alfred Wallace headed for the upper Rio Negro, 'looking forward with hope and expectation to the distant and little-known regions' he was about to visit.

Wallace travelled initially with a Portuguese, Senhor Lima, an old-established trader. The canoe was 'tolerably roomy', about thirty-five feet long and seven broad: it was heavily loaded with trading goods, six months' supplies for Lima's family, and Wallace's own belongings and collecting gear. For days, weeks, they rowed (or rather were rowed) upriver – it comes as something of a surprise when Wallace notes, on 30 September, that they 'again saw the opposite side of the river, and crossed over where it is about four miles wide'. Another twelve days followed, by which time the stream was too strong for them to make much headway, and they transferred to two smaller canoes to take them up the series of rapids, the Falls of the Rio Negro. In the heart of these was the village of São Gabriel, equipped with a fort and a *comandante*, whose permission was required before anyone could continue upriver. Another couple of days, and they entered the 'great and unknown "River Uaupés"', and so reached the village where Lima lived. Wallace, though having reservations about the way Lima conducted his affairs – 'Senhor L had informed me during the voyage that he did not patronise marriage, and thought everybody a great fool who did' – was very much reliant on him for the services of his Indian hunters. But he soon grew tired of the limited opportunities around the village, though he managed to shoot some birds every day, and arranged an expedition into the forest in search of the *gallo*, the 'Cock of the Rock'. He moved on from village to village up a narrow stream; at midnight they drove stakes into the sandy ground to hang their hammocks. The Indian boys shot birds for him with their blow-pipes, or crept along by his side and silently pointed out birds or small animals before he had seen them. Left to his own, Wallace began to communicate with the Indians on his own terms.

> By the promise of good payment for every 'Gallo' they killed for me, I persuaded almost the whole male population of the village to accompany me. As our path was through a dense forest for ten miles, we could not load ourselves with much baggage: every man had to carry his gravatana, bow and arrows, rede, and some farinha; which, with salt, was all the provisions we took, trusting to the forest for meat; and I even gave up my daily and only luxury of coffee.

Even here Wallace was not the first scientific traveller to arrive from Europe – at the last house on the road to the Serra he saw a young *mamelúca*, 'very fair and handsome, and of a particularly intelligent expression and countenance' – evidently the daughter of the celebrated German naturalist Dr Natterer: 'She was a fine specimen of the noble race produced by the mixture of the Saxon and Indian blood.'

The path into the forest was at first tolerable:

> Soon, however, it was a mere track a few inches wide, winding among thorny creepers, and over deep beds of decaying leaves. Gigantic buttress trees, tall fluted stems, strange palms, and elegant tree-ferns were abundant on every side, and many persons may suppose that our walk must necessarily have been a delightful one; but there were many disagreeables. Hard roots rose up in ridges along our path, swamp and mud alternated with quartz pebbles and rotten leaves; and as I floundered along in the barefooted enjoyment of these, some overhanging bough would knock the cap from my head or the gun from my hand; or the hooked spines of the climbing palms would catch in my shirt-sleeves, and oblige me either to halt and deliberately unhook

> myself, or leave a portion of my unlucky garment behind. The Indians were all naked, or, if they had a shirt or trousers, carried them in a bundle on their heads, and I have no doubt looked upon me as a good illustration of the uselessness and bad consequences of wearing clothes upon a forest journey.

At the foot of the Serra, the party of twelve ran into a herd of wild hogs, and scattered. That night, in a large cave by a stream, there was a feast of stewed pig. Each day, Wallace and his hunters clambered over the precipitous mountain in search of the *gallo*; and each evening, after skinning the day's catch, they sat round the fire: 'The fires were made up, the pork put to smoke over them, and around me were thirteen naked Indians, talking in unknown tongues.' Two could speak a little Portuguese, and cross-questioned Wallace about where iron came from, how calico was made, if paper grew in his country – they were astonished to hear that everyone there was white, and could not imagine how white men could work, or how there could be a country without forest. After nine days, living off monkeys and birds as well as the wild pig, they made their way back to the village. Wallace was pleased with his collection: twelve *gallos*, two fine trogons, blue-capped manakins, barbets, and ant-thrushes. More importantly, he had proved that he could exist happily on his own terms, or perhaps on their terms, with the Indians, whom he clearly respected much more than their Portuguese 'masters'. He was planning a new expedition, but the Indians would not leave until the itinerant padre had arrived, a tall, thin, prematurely aged man 'thoroughly worn out by every kind of debauchery' – Frei Jozé dos Santos Innocentos claimed that he never did anything disreputable during the *day*. The next morning, there were lots of baptisms and a few weddings, and Frei Jozé gave a practical homily on the duties of the married state, which, as Wallace commented, might have done some good if the married couples had been able to understand it: the only two other white men present, besides Wallace, were conspicuously unmarried with large families.

In January 1851, with four Indians and a canoe lent to him by Lima, Wallace headed upriver, travelling light: watch, sextant and compass, insect- and bird-boxes, gun and ammunition, salt, beads, fish-hooks and calico for the Indians. They were heading for São Carlos, the principal Venezuelan village on the Rio Negro, the furthest point reached by Humboldt from the opposite direction fifty years before. Wallace, by now reasonably fluent in Portuguese, found himself grappling with Spanish. He was at the point where the Casiquiarì, which links the Orinoco and Amazon systems, joins the Rio Negro. Partly by river and partly by land,

Wallace explored the division between the two river basins. Once, walking quietly in the forest on his own, he saw a large jet-black animal emerging on to the path twenty yards ahead. 'As it moved slowly on, and its whole body and long curving tail came into full view in the middle of the road, I saw that it was a fine black jaguar.' The collector took control of the naturalist for a moment, and he raised his gun to his shoulder; then, remembering that both barrels were loaded with small shot, he thought better of it, and stood silently gazing: 'In the middle of the road he turned his head, and for an instant paused and gazed at me, but having, I suppose, other business to attend to, walked steadily on, and disappeared in the thicket.' This forest meeting pleased Wallace: he had seen, in its native wilds, a magnificent specimen of the most powerful and dangerous animal of the American continent. But one such encounter was enough for that day; as it was nearly sunset, he turned back towards the village.

Wallace made the village of Javíta his base, but luck was against him as far as the weather was concerned. On the very night he reached the village the rains began, and continued daily each afternoon and night, making collecting difficult, and the business of drying, preserving and cataloguing laborious. Even so, it was a rich ground for a naturalist. In the six weeks or so that he was there, he found forty species of butterfly quite new to him, including an abundance of the great blue butterflies, Morpho Melenaus and Morpho Helenor; in the forest he saw monkeys, agoutis, coatis, many beautiful trogons, and numerous snakes; the Indians brought him fish, which served for supper as well as science, and a curious little alligator, which, to their amusement, he skinned and stuffed. Working each afternoon, he was exposed to the sand-flies, and by the time he had finished his hands and wrists were as rough and red as a boiled lobster; but after they were bathed in cold water, the inflammation subsided. In spite of these frustrations, he was extremely content, and even wrote a long poem inveighing against civilised life – though, as he admits, he wrote it partly to relieve the monotony of his situation:

There is an Indian village; all around,
The dark, eternal, boundless forest spreads
Its varied foliage. Stately palm-trees rise
On every side, and numerous trees unknown
Save by strange names uncouth to English ears.
Here I dwelt awhile the one white man
Among perhaps two hundred living souls.
They pass a peaceful and contented life,
These black-hair'd, red-skinn'd, handsome, half-wild men . . .

The women and young girls were 'far superior' in their graceful forms to English village maids:

> For their free growth no straps or bands impede,
> But simple food, free air, and daily baths
> And exercise, give all that Nature asks
> To mould a beautiful and healthy frame.

The boys ran and raced and shouted and leaped; and the forest and the river provided all the village's wants. Wallace did not place the civilised below the 'savage'; he was too conscious of the lack of intellectual pleasures and delights 'that the well-cultivated mind enjoys'; but he could not help comparing the simple, peaceful life of the Indians with the existence of the urban, and the rural, poor and oppressed of Europe:

> For are there not, confined in our dense towns,
> And scattered over our most fertile fields,
> Millions of men who live a lower life –
> Lower in physical and moral health –
> Than the Red Indian of these trackless wilds?

The poem is dated March 1851, and Wallace commented that he wrote it 'in a state of excited indignation against civilised life in general', and 'not altogether as my views when writing in London in 1853'. Nevertheless, it came out of first-hand experience, from a man sharing many of the living conditions of the people he described: the noble savage of Rousseau could and did exist.

> I'd be an Indian here, and live content
> To fish, and hunt, and paddle my canoe,
> And see my children grow, like young wild fawns,
> In health of body and in peace of mind,
> Rich without wealth, and happy without gold!

In the same year that Wallace's account was published, Charles Dickens wrote his essay 'The Noble Savage', in which he defined a savage as 'a something highly desirable to be civilised off the face of the earth'.

Soon after putting his meditations into verse, Wallace woke one morning to find that his Indian helpers had vanished; and he made his way back by degrees to Guía, where he was soon planning a fresh expedition up the River Uaupés, this time in Lima's company. This journey took him even further into little-known territory – he found himself in the presence of 'the true denizens of the forest', so different from the half-civilised races among whom he had been living that it was as if he had been suddenly transported to a different country.

> Some two hundred men, women, and children were scattered about the house lying in the maquieras, squatting on the ground, or sitting on the small painted stools, which are made only by the inhabitants of this river. Almost all were naked and painted, and wearing their various feathers and other ornaments. Some were walking or conversing, and others were dancing, or playing small fifes and whistles. The regular festa had broken up that morning; the chiefs and principal men had put off their feather head-dresses, but as caxirí still remained, the young men and women continued dancing. . . . The men and boys appropriated all the ornaments, thus reversing the custom of civilised countries and imitating nature, who invariably decorates the male sex with the most brilliant colours and most remarkable ornaments.

Wallace, as he would do in the Malaysian Archipelago, becomes anthropologist as much as naturalist; he attended another, regular *festa*, and experimented by emptying a calabash of *caxirí* – exceedingly good, he reported, even though the *mandiocca* cake of which it was made was chewed 'by a parcel of old women'. However, he only observed the effects of the powerful narcotic *caapí* (unlike Richard Spruce on another occasion).

Wallace had been planning to travel west, to the Andes. But he was tempted by what he had seen to return instead to the Uaupés, equipped this time to make a collection of live birds and animals. To do this meant, first, returning to Barra, to supervise the dispatch of his collections, and purchase another load of goods for barter – a mere round trip of 1,500 miles. There he found Spruce, 'a prisoner for want of men', and lodged with him. From Pará, there was gloomy news: a letter, three months old, informing him that his brother Herbert was dangerously ill with yellow fever. Wallace assumed the worst, rightly as it turned out, though he found it strange that no one had written subsequently to confirm the death. But nevertheless he threw himself into preparations for his return to the Uaupés, even having to make his own insect-boxes and packing cases, in the absence of the only carpenter in the town. He and Spruce wished they could travel together, but Wallace's canoe was too small for them both and he did not have enough men to manage Spruce's larger vessel. So Spruce accompanied him in a small *montaria* (canoe) for a day, before turning back.

Although Wallace penetrated further up the Uaupés than any other previous European traveller, one senses that he was reaching the end of his span of energy. Perhaps the realisation of his brother's probable death was affecting him. His own health had become less reliable: he suffered several bouts of dysentery and fever, one especially severe. Two of his quarries, the painted turtle and a rumoured white umbrella bird, eluded him. He began to build up a collection of live animals and birds, mostly monkeys and parrots, which took a good deal of looking after. He brought them downriver, stopping off to see Spruce at São Gabriel on the way. At Barra, he picked up four large cases of specimens which should have been shipped to England a year before, but which had been held up by customs regulations; and received confirmation of his brother's death. By the time he left Barra for the coast, the hundred or so creatures he had acquired had dwindled to thirty-four: five monkeys, two macaws, twenty parrots or paroquets, a white-crested pheasant, some small birds, and his favourite, a full-grown and very tame toucan. The night before they left for Pará, the toucan flew overboard, and was drowned. It was a bad omen. On 12 July 1852, still weak with fever, Wallace embarked for London on the *Helen*.

Three weeks into the voyage, Wallace was reading in his cabin when the captain came to tell him the ship was on fire: the cargo, which included balsam, was smouldering. Chaotic preparations for taking to the boats followed: 'The cook was sent for corks to plug the holes in the bottom of the boats. Now no one knew where a rudder had been put away; now the thowl-pins were missing.' Wallace returned to his smoke-filled cabin and, in a kind of daze, only had the will to salvage two notebooks and some drawings, leaving his books and instruments behind. He took to the boats with the rest of the crew and, as they watched the burning ship, he could see the remaining monkeys and parrots retreating to the bowsprit; but only one parrot could be coaxed off the ship: it fell in the water, and was scooped out, scorched and bedraggled.

They were in the boats for ten days, before being picked up by the *Jordeson*, sailing from Cuba. Even that was not the end of the misfortunes; twice, the new ship almost ran out of provisions, and a final storm in the Channel left four feet of water in the hold. On 1 October, after a voyage of eighty days, Wallace landed at Deal, conventionally 'thankful for having escaped so many dangers, and glad to tread once more on English ground'. The one piece of good fortune, or foresight, was that Stevens, the agent, had on his own initiative insured Wallace's cargo for £200. Financially, at least, he had not lost so much as he feared. But the lack of specimens was a terrible blow: 'With what pleasure had I looked upon every rare and curious insect I had added to my collection! How many times, when almost overcome by the ague, had I crawled into the forest and been rewarded by some unknown and beautiful species! How many places, which no European foot but my own had trodden, would have been recalled to my memory by the rare birds and insects they had furnished to my collection!' The expedition which might have set Wallace up for life, by furnishing him with the materials for future research, had been ruined when it was all but accomplished. But he did not wait for long before plunging into the life of scientific London. Three days later, he was at a meeting of the Entomological Society, even though, as he wrote to Spruce, he could scarcely walk on his swollen ankles, 'my legs not being able to stand work after such a long rest in the ship'.

Wallace attended these meetings, not as a member, but as a visitor. Slowly, very slowly, his status changed from that of a mere collector to someone whose views were listened to with respect. He gave papers to the Entomological and Zoological Societies, and, on 13 June 1853, one on the Rio Negro to the Royal Geographical Society, where he met the president, Sir Roderick Murchison. Wallace already knew he must make another expedition, whether to the Andes or to Asia. Soon he was making an official application for the Council of the Royal Geographical Society to endorse his project, and to recommend he should be given a free passage. The Amazon trip had confirmed his potential, even if he had been denied its full fruits. As he expressed it in his submission, 'As some guarantee of his capabilities as a traveller he may perhaps be excused for referring to his recent travels for nearly five years, in South America, where alone and unassisted he penetrated several hundred miles beyond any former European traveller.'

9. Alfred Russel Wallace in 1869, aged forty-six.

4 From the Amazon to the Andes

As Wallace and Bates were exploring and collecting further inland up the course of the Amazon, a third English naturalist, Richard Spruce, sailed from Liverpool on 7 June 1849. With him were an assistant and companion, Robert King, and Alfred Wallace's younger brother, Herbert.

Spruce was one of the most autonomous and modest of Victorian scientists. He had no family connections, no inherited wealth to ease his way in the world. He brought himself to the attention of influential men such as William Hooker and George Bentham by his brilliance, and his meticulousness, as a naturalist. He never enjoyed robust health, though he felt much better in the open air than in the cold stone classrooms and schoolhouses of Yorkshire; even so he survived fourteen consecutive years in the Amazon basin and in the Andes, living for much of the time beyond the reach of Western medicine. As a collector, he was supreme. Yet he was interested not just in the plants he especially loved, the mosses and hepaticae (liverworts), but in the forest trees: in fact, he was absorbed by every aspect of the new life he encountered. He learned to speak and write Portuguese and Spanish, and some seven languages of the Indians. He recorded the Indians' words, their customs, their medicine, their food, writing down everything in his beautifully even and legible script in a calm, methodical way, however awkward and uncomfortable his surroundings.

Spruce was born at Ganthorpe, near Castle Howard, in Yorkshire, the son of a schoolmaster. At sixteen, following the tradition of Gilbert White of Selborne, he had made a collection of 403 local plants, and, building on this early interest, he began to hunt further afield, specialising in mosses and hepaticae. He published accounts of his discoveries in the *Phytologist*, and this brought him into correspondence, and later friendship, with other collectors and specialists such as Hooker. His mother died when he was very young, his father married again and produced eight daughters, and Spruce, with only his excellent mathematical and classical education and his self-taught expertise as a naturalist to draw on, turned to teaching to earn a living. He found the routine taxing and uncongenial, and the indoor life seemed bad for his health. When the York school he was employed by closed down, he was offered another post at a boarding-school, but rejected it. The only alternative was a botanical job, and William Hooker put before him the possibility of a post as curator of some colonial botanical garden: Hooker, recently appointed director at Kew, was beginning to establish his worldwide network. Then Hooker proposed another scheme: Spruce should go to the French Pyrenees – Spain was eliminated as being too unsettled and dangerous – and pay for his trip by selling sets of plants. He left in April 1845, and was away for almost a year. He had a wonderful time, climbing up to 10,000 feet in search of flowers, and rummaging through the beech and elm forests for mosses. He identified seventeen completely new mosses, and located seventy-three more which had never previously been found in the Pyrenees. Between 300 and 400 sets of plants had to be named, arranged and distributed to his subscribers; and then he was able to settle down to write up his speciality, *The Musci and Hepaticae of the Pyrenees.*

Back once more in Yorkshire, Spruce grew restless. His father became ill, and for two months Spruce filled in for him at the school on the Castle Howard estate. In the autumn of 1848 he was in London to supervise the sale of a friend's herbarium. There, at the British Museum and at Kew, he heard about the collecting successes of Bates and Wallace, and (ironically, in view of what was to come) of how highly they spoke of the climate. He met George Bentham, who encouraged him and generously offered to act as a sorting-office and distributor. Hooker urged him on. The Earl of Carlisle at Castle Howard gave him support. He obtained the precarious security of eleven firm subscribers for his specimens: this was later increased to thirty, as the value and precision of his collecting became established. Financed solely by the sale of specimens which he would export back to England, he embarked on an expedition which would extend for an astonishing fourteen years.

Spruce made the first leg of his voyage up the Amazon in comparative comfort, taking a passage for the 474 miles from Pará up to Santarém on Captain Hislop's brig, the *Três de Junho*. Hislop, a 'sturdy, rosy Scotsman', had been on the Amazon for forty-five years, and had acquired a number of mildly eccentric habits. He kept files and files of newspapers, preferring them at least six months old, and read two books only: Volney's *Ruins of Empires*, and the Bible; after a few extra glasses of port, Hislop would expand on the character of Moses, whom he described as 'a great general, and a great lawgiver, but a great impostor'. Hislop found Spruce a house, provided a few chairs and planks for shelves – vital for a collector – and lent him one of his slaves as a temporary cook. A few days later Alfred Wallace arrived, to be reunited with his brother, and he was soon showing Spruce the paths across the campo. They took a keen pleasure in their evening conversations though they found it hard to keep awake after eight in the evening – Spruce had not yet adjusted to the tropics by taking a siesta.

But Spruce did not stay long in the comparative luxury of Santarém. He decided first to explore the River Trombetas, from Obydos. The military commander lent him a canoe, and coerced three rather reluctant Indians to do the paddling and navigation. This was a kind of test journey, an initiation into the dos and don'ts of forest travel. King, Spruce's assistant, had a severe fright – he had slept in the open on a sandbank, only to find the next morning the tracks of an alligator passing rather too close for comfort. They then got themselves thoroughly lost, after pushing up a tributary stream. Two of the Indians disappeared, without a word; and King somehow became separated from Spruce and the remaining Indian during a violent rainstorm. It was another three hours before he found them – in the confusion he had gone up rather than downstream. Spruce then made a second mistake – they reached the main river, which gave him a sense of security, so he sent Manoel, the remaining guide, ahead, to light a fire and begin cooking. Night fell, and the moon had not yet risen above the treetops: 'We sat down at the foot of a large tree, in the angle between two sapopemas; but both tree and ground were very wet, and we ourselves were thoroughly soaked, for, even after the rain ceased, every bush we pushed through, every liana we cut, brought down on us a shower of drops.' The situation, Spruce comments laconically, 'was no enviable one': their only arms were King's machete and his own lichenological hammer; they had no materials to light a fire. 'We had in a bag a little roast pirarucú (fish) and farinha, and although the latter had been transformed into a glutinous paste by the rain, we made a scanty meal on them. After a while we began to feel chilly and drowsy; but had we given way to sleep under such circumstances, we might have awaked too stiff to move; to say nothing of the risk of being assaulted by jaguars . . .' They scrambled on, plunging into prickly palms, getting entangled in equally prickly sipós: 'One's foot trips in a trailing sipó – attempting to withdraw it, one gives the sipó an additional turn, and is perhaps thrown down; or, in stooping to disentangle it, one's chin is caught as in a halter by a stout twisted sipó hanging between two trees. At one time we got on the track of large ants, which crowded on our legs and feet and stung us terribly.' They stumbled into camp at one o'clock in the morning. The effects lingered with them a full week: rheumatic pains, stiffness, ulcers from the wounds in their hands, feet and legs. Spruce learned a great deal in a short time: later in his travels he would make light of heart-stopping combinations of climate, illness and privation. For the moment, he was relieved to return to Santarém, which he made his base for the next twelve months.

After exhausting the flora available within the space of a day's journey from Santarém, Spruce planned to move to the mouth of the Rio Negro, at Barra. This was not so easy to accomplish. He needed a crew to accompany him, and most of the available 'free men of colour' were in debt to the Santarém traders, who would not allow them to leave without settling their accounts. Eventually, he set off with three crew in a small canoe; his baggage half filled it, and the palm-leaf cabin leaked, spoiling his store of paper – one of Spruce's main collecting problems was keeping the paper for his specimens sufficiently dry.

Spruce was a little frustrated by being confined to a canoe in mid-stream, but as always made the most of his circumstances. He could not do much collecting, though twice he swam ashore to gather 'a stout Mimoseous twiner'; further upstream, there were too many alligators. Even in the enforced frustration of a slow voyage, his curiosity about the environment was alert; he listened and observed, and learned from his Indian crew:

> Rarely is there perfect silence on the banks of the Amazon. Even in the heat of the day, from 12 to 3 o'clock, when birds and beasts hide themselves in the recesses of the forest, there is still the hum of busy bees and gaily-coloured flies, culling sweets from flowering trees that line the shore, especially from certain Ingas and allied trees; and with fading twilight (6½ p.m.) innumerable frogs in the shallows and among the tall grasses chaunt forth their Ave Marias, sometimes simulating the chirping of birds, at others the hallooing of crowds of people in a distant wood. About the same hour the carapaná (mosquito) begins its night-enduring song, more annoying to the wearied voyager than even the wound it inflicts. There are, besides, various birds which sing, at intervals, the night through, and whose names are uniformly framed in imitation of their note; such are the acurau, the murucututú – a sort of owl – and the jacurutú, whose song is peculiarly lugubrious. A sort of pigeon, which is heard at 5 o'clock in the morning, is called, and is supposed to say, 'Maria, ja he dia!' ('Mary, it is already day!') – a name which reminded

one of 'Milk the cow clean, Katey!' a Yorkshire appellation of the stockdove.

He amused his companions by asking what bird was making a croaking noise in a cacao tree – it turned out to be a rat-sized animal, whose replies were much valued by Indian *payes*, 'or wizards'. He recorded one of his crew's conversations:

'Your worship sings very sweetly all alone by night in the cacao tree!' – '*Torô! Torô!*'

'Your worship seems to be enjoying your supper on the delicious cacao!' – '*Torô! Torô!*'

'Will your worship tell me if we are to have a favourable wind in the morning?' – Toro respondeth not.

'Your worship, do me the favour to say if we shall arrive at Obidos tomorrow?' – Again no reply.

'Your worship may go to the devil!' – An insult of which Torô taketh not the least notice; and so ends the dialogue, the Indian being too angry to interrogate further.

Spruce and King reached Barra on 10 December; the 400 miles had taken them sixty-three days. Spruce took his letters of credit to Senhor Henrique, known as the travellers' friend, who installed him in a new two-storey house and invited him to dinner. For almost a year he made the town – soon to grow into a city – his base. In his first eighteen months on the Lower Amazon, he had collected over 1,100 species of plants; at the mouth of Rio Negro, he added another 750 species, visiting every area around, and making short expeditions up the larger streams which joined the main river in the locality. All the time he was learning new skills, as on a visit to one of Senhor Henrique's farms:

Not a day passed without rain. Sometimes there was sun enough in the morning to enable me to dry my paper before setting out to herborise. When there was not I took the paper across the river in the evening and got it dried on the forno. This is a narrow rapid stream winding through dark forests, the climbers of which often stretch across it and are troublesome to avoid as the canoe shoots beneath them. The first time I made the passage, along with my attendant Pedro, he placed himself in the prow and I in the poop of the canoe, each of us with a paddle; but although I was well accustomed to steer by means of a rudder, I had never attempted it with a paddle, and my want of skill brought us up every now and then plump into the bushes, which I could see ruffled Pedro's equanimity no little. After we landed, I heard him say to his sister in Lingoa Geral, 'This man

> knows nothing - I doubt he could even shoot a bird with an arrow!' (a feat which every boy of twelve is supposed capable of performing).

Spruce consoled his wounded pride by the thought that the most eminent botanists of Europe would have done little better. Most of them would have done worse. Spruce, partly because of his lack of money, partly because of his character, was a hands-on traveller, at least until his health deteriorated. He also learned how to get the best out of his Indian companions, something he felt he did better than his great forerunner, Humboldt. 'The Indians do well enough in the field when one knows how to manage them. . . . It does not do to ask them to do anything *as a task*, however much money, etc., you may offer for the performance of it. My usual invitation is "Yasso yaoata" ("Let us go for a walk"). We get into our montaria (canoe), enter one of the igarapes (small streams), and when we reach the heart of the forest they are all alacrity to climb or cut down the trees, the gathering of the flowers being all the while represented as a mere matter of amusement.' On another trip, he found himself at Manaquiry during the feast for the Vesper of St John. Spruce was led forward and invited to open the dancing with the queen of the feast, the 'Juiza'. 'I saw that it was intended to do me honour and that I should be accounted very proud if I refused. I therefore led the lady out, first casting off my coat and shoes in order to be on terms of equality with the rest of the performers. We got through the dance triumphantly, and at its close there was a general viva and clapping of hands for "the good white man who did not despise other people's customs!"' Once 'in for it', Spruce joined in the celebrations all night, dancing with every girl in the room, though sitting out the more physical Indian dances – the songs accompanying them, he comments, were in the Lingoa Geral of the Indians, and 'of such a nature as not to admit of their being decently translated into any European language'. At daybreak everyone swam, and then began to prepare for breakfast, slaughtering a pig and a turtle; but Spruce did not want to waste a day's work, so went back to his plant-hunting.

While collecting hard during this year, and writing long and detailed letters both to his friends and to his botanical contacts – Sir William Hooker, George Bentham, John Smith the curator at Kew – Spruce was also making preparations for his expedition to the Orinoco, acquiring and fitting up his own canoe and negotiating for a crew. This was a difficult business: it was best to get crew members from your destination, writing ahead to some contact, perhaps a local governor, who would arrange to send a group of men down to your starting place – there was then less chance of the crew slipping quietly away home during the journey. He also spent days buying trade goods: money was useless, and he had to lay out a 'whole fortune' in prints and cotton fabrics, axes, cutlasses, fish-hooks, beads and looking-glasses. On 14 November 1851, he set off.

'This day (Friday) I left the Barra in my canoe with six men, for the Upper Rio Negro. There was little wind, which soon failed entirely. We slept at Paricatuba, about fifteen miles from the Barra on the opposite shore, where I gathered seeds of a beautiful small tree allied to the well-known *Lagerstroemia indica* of our conservatories.' For the first time, in his own canoe, he was the master of his movements, could stop when he liked and go on when he liked. He described the cabin to his Yorkshire friend John Teasdale:

> It was long enough to suspend my hammock within it, and I made myself besides a nice soft bed of thick layers of the bark of the Brazil-nut tree (which you will find mentioned by Humboldt under the name of Bertholletia); my large boxes ranged along the sides served for tables and the smaller ones for seats; while from the roof I suspended my gun and various things that I required to have constantly at hand. The fore-cabin or tolda da proa was occupied by baskets of farinha, a few bushels of salt, and various other things which I was taking with me to barter with the Indians; it served also as a sleeping-place for the men when the weather was wet, otherwise they preferred sleeping outside.

As for Spruce, warned by past experience of fever, he always slept inside; though on the 'black waters' of this river there were no *carapanás*, so he could pass the evening and the early part of the night outside, under the moon; there was, however, a full complement of other biting insects during the day.

In terms of plant collecting, the journey was a great success. The Indians would keep a lookout during the hot afternoons, when Spruce was busy with his papers in the cabin – 'O patrão! aikué potéra poranga' ('Master! here's a pretty flower'); and he would turn out to see if it was something new, as it often was. But as they ascended the river the going got harder, with falls and cataracts which were time-consuming to traverse. And at São Gabriel he found that the bulk of the inhabitants was the garrison of fourteen: the army of Brazil in the frontier posts was recruited from criminals, so that half at least were murderers: more crucially, they broke into the house, and raided the supplies. Instead of his snug cabin in the canoe, he had to move into an old house whose thatch was stocked with rats, scorpions, cockroaches and vampire bats, and an earth floor undermined by sauba ants, who attacked his supplies of *farinha* and, worse, his dried plants. He burnt the ants, smoked them, drowned them, and trod on them; but they required constant vigilance. Meanwhile, the termites were munching their way into a fortunately empty packing-case. It was impossible to buy an egg or even a banana, but luckily Spruce had two Indians with him in the house, a hunter and a fisherman.

He wanted to explore the granite serras which rise around São Gabriel: 'In a caatinga at the base of the Serra de Gama I made an interesting collection. There are also other caatingas or "white forests" in the neighbourhood: the soil a thin covering of white sand over granite, the trees low, twiners scarcely any, trunks hung with Ferns and Orchises, branches with Hepaticae . . .'. However, as the year went on, conditions grew more unpleasant. His hunter fell ill. The *festa* began on the eve of the Ascension, and lasted a month, during which time no one would hunt or fish; and the rains came, so that the river rose and made fishing impossible. Spruce went out himself with his gun, hunting for parrots; for

10. Rocks in the Cataracts of São Gabriel. (Drawing by Richard Spruce)

three days he had nothing but *xibe – farinha* mixed with water; and when someone like Spruce says he was never so near dying of hunger, one is inclined to believe him. As he wrote to his mentor George Bentham in August, 'My last dates from England are a year old. Neither newspapers nor anything else ever reach me now. I seem to have taken my last leave of civilisation'.

Spruce decided to make a trip up the Uaupés river, where Wallace had recently been. He chose Panuré as a base camp because there were some European families there temporarily, canoe building, who would look after his things when he was plant-hunting. This proved an excellent area. He found 500 or so species in the locality, of which he estimated 400 were entirely undescribed. In fact, he made himself ill by overworking in his enthusiasm. Once absorbed in a new discovery, very little distracted him. He had a tame agamí, or trumpeter, a bird which is a fearless snake-hunter. It became so attached to Spruce that it followed him about like a dog, never failing to kill any snake they met:

> One day I was alone with the agamí in a caatinga about four miles from the village, where I lingered about a good while in a spot comparatively clear of underwood, but abounding in certain minute plants (Burmanniaceae) which I was much interested to gather. Whilst I hunted for plants the agamí hunted for snakes, and had already caught three or four, which it brought and laid before me as it caught them. I suppose I had not noticed and praised its prowess as I usually did, for at length – apparently determined to attract my attention – it laid a newly-caught snake on my naked feet, when I was standing erect, absorbed in the examination of a little Burmannia with my lens. The snake was scarcely injured, and immediately twined up my leg. To snatch it off and jerk it away into the bush was the work of a moment; but ever afterwards I took care to leave the agamí at home when I started for the forest.

Spruce thought the agamí might with advantage be introduced to British India, where it was likely to prove a superior snake-hunter to the mongoose.

The following nine months proved especially difficult. To the relentless isolation which Bates had suffered were added physical danger and serious illness. He moved again, upriver, towards the border of Brazil and on into Venezuela, from Portuguese to Spanish influence. He found collecting arduous: the Indians were lethargic, and too often drunk. Food was extremely scarce: his time was taken up 'in procuring materials for a miserable existence': 'I write now under most unpleasant circumstances,' he told Bentham from San Carlos on the Rio Uaupés, 'and God only knows whether I shall live to close this letter.' There was a rumour that a massacre of the whites was planned for the feast of St John: 'Some said that they had submitted long enough to the whites, and that on the Orinoco it was quite a common thing to kill a white man and throw his body into the river.' Spruce, the least aggressive of men but extremely stubborn, prepared to fight for survival. He and the two young Portuguese men pooled their weapons – seven firearms, two swords and a heap of cutlasses – and sat out a long night, and several apprehensive days after. Threats were made, muskets fired and drums beaten; but no attack came. Spruce, stuck in this uncomfortable spot, did as much collecting as he could, and did find some interesting mosses and hepaticae; but his mind was turning towards the next stage of his travels: his voyage up the Casiquiari to Esmeralda on the Orinoco, a voyage which began in the footsteps of Humboldt and Bonpland, but which continued up two rivers never before explored by a European.

Meanwhile, he caught up on his correspondence, packed up consignments of plants for Kew, and wrote a grateful letter to the Countess of Carlisle to accompany his gift to Lord Carlisle of an Indian hammock, as some kind of acknowledgement of the help he had received before he set out on his South American travels. It had been commissioned nine months before, and had taken an Indian girl four months to weave. It was, he admitted, a substitute for the living plants which he had hoped to send to the Castle Howard conservatories: 'On the 12th of this month I shall have been four years in South America. During the whole of this time I have been in the midst of forest – lofty, dense and at first glance seemingly impenetrable. The trees composing it are generally from their size alone unfitted for cultivation in our conservatories, and the few whose dimensions would not exclude them, are mostly species bearing small and obscurely coloured flowers. The seeds are so frequently resinous or oily as to lose their vitality when kept long out of the earth'; besides, the sending of living plants was hazardous and tedious. The hammock duly took its place in the Castle Howard museum, a strange artefact from the heart of the Amazonian forest to be housed in Vanbrugh's neo-classical masterpiece.

Spruce's visit to Esmeralda was a deliberate act of homage to Humboldt, who had described it in his *Narrative*. The village, now reduced to six miserable huts, had the most magnificent site Spruce had seen in South America.

> Between the Cerro Duida on the west and the mountains of the Guapo and Padamo on the east extend wide grassy savannas in which almost the only trees are scattered fan-palms (Moriches). On the side next the Orinoco a semi-circular ridge of fantastically-piled granite blocks, in whose crevices grow a few scattered shrubs, cut off a small savanna on which stands Esmeralda. All up and down the Orinoco, and on the margins of the savanna, rise hills of granite and schist, some nearly naked, others forest-clad, and at the back (to the N.W.) rises the abrupt and frowning mass of Duida. If you can fancy all this seen by a setting sun – the deep ravines that furrow Duida on the east buried in nocturnal gloom, while the salient edges glitter like silver (the rock is chiefly micaceous schist) – you will realise in some degree a scene which has few equals.

But if to the sight Esmeralda was a Paradise, in reality it was an Inferno, scarcely habitable by man. There were no birds, not even a butterfly, to be seen; amid the luxuriance of vegetable life, animal life was almost extinct. But this absence was only apparent: 'If I passed my hand across my face I brought it away covered with blood and with the crushed bodies of gorged mosquitoes.' The few human inhabitants drowsed away the day inside their huts like bats, only stealing out at dawn and dusk to look for food; and when

11. Cerro Duida (8,000 feet), from the Cross near the Village of Esmeralda. (Spruce, December 1853)

Spruce came back from one of his herborising walks, his hands and feet and neck and face were covered with blood. He had to eat walking about, on food always 'well peppered' with mosquitoes. Not surprisingly, he found working on his plants very difficult, even though he put on gloves and tied his trousers down over his ankles.

As soon as he could, he moved out to explore the Cunucunúma. But the river level was low, and falling; and after exhausting attempts to drag the canoe up a series of falls, Spruce had to abandon his plan 'with a sorrowful heart'. 'I had calculated on spending at least a month among the Maquiritares and exploring their river by means of small boats up to its sources . . . but this was impracticable unless I could get my stock of paper and goods to some station which I could make my headquarters'; in the lower reaches of the river, the forest was so dense it was even hard to find a spot of ground to cook on. Instead, he turned anthropologist for a few days. He clearly had particularly good relationships with the Maquiritares. He gives detailed descriptions of them, and some of their customs, and especially of one of their chiefs, Tussarí, 'a remarkable man', with whom he stayed. He did manage to work his way up the Pacimoni river, and spent a month there collecting. He returned to San Carlos, sorted out his collections, and then set out on another expedition to the cataracts of Maypures.

Ascending the river further to a village called San Fernando, Spruce fell seriously ill. It had been a four-day journey; for the last two, he was helpless with fever. His stock of dried plants – he had collected successfully around Maypures – took up most of the small cabin, so he was forced to lie half outside, exposed day and night to sun and rain. He arranged to be carried to the house of an elderly woman, a Zamba called Carmen Reja, who agreed to nurse him; it was an unhappy arrangement. Spruce suffered violent attacks of fever at night; he vomited painfully, had an unquenchable thirst, could take no food, and had great difficulty in breathing. He gave instructions to the local Comisario as to what to do with his plants and belongings, and waited for death in a state of apathy. Meanwhile, his nurse, according to him, in recollections certainly heightened by his near delirium, either abandoned him for hours at a time or filled the house with her friends, to discuss and abuse him. 'Die, you English dog!', she called out, 'that we may have a merry watch-night with your dollars!' Spruce heard one of her companions whisper she might have to give the white man some poison, to finish him off. After nineteen days the fever left him, but it was a further two weeks before he was strong enough to escape from this nightmare in the company of a Portuguese trader, and even then he had to be carried in a hammock to the village where he had left his larger canoe.

He finally set off from San Carlos on the last leg of his trip back to Barra at the end of November. For the first time, he came under serious threat from his Indian crew. He had slung his hammock in a small shed, and turned in to sleep after an evening meal of forequarter of alligator. His men were drinking, and as the conversation grew louder, Spruce realised it was all about him. (They probably could not accept how fluent he had become in their language, or else were too drunk for discretion.) The first plan was simply to abandon him, after being paid in advance. Then one of them began to speculate about 'the man's' merchandise, not realising it was mostly paper and plants: they would have to kill him on the way. From there, they progressed to deciding to do the job at once: 'We have him here sleeping in the midst of the forest, far removed from all observers. When he left San Carlos every one knew him for a sick man, and no one will be surprised to hear of his death.' All evening Spruce had been making regular trips to the forest to relieve himself, as he was suffering from diarrhoea. Just as he lay down, he heard them decide that the best solution was to strangle him once he was asleep again. He kept his feet on the ground, ready to spring up; when he heard a whisper 'Now it is good!', he got up, walked casually towards the forest as before, then turned abruptly and climbed into the canoe. There he unlocked the cabin, barricaded the entrance with a mound of paper, and rested his double-barrelled gun on it. He passed the rest of the long night on guard. When day came, he managed to dispense with the services of the ringleader and set off calmly, but watchfully, down the Uaupés. As he commented to Hooker, for five years he had placed the utmost confidence in the Indians, sleeping unarmed, strolling off into the forest when they halted to cook, when they could easily have abandoned him; this time, he had to adopt an entirely different method, and had to keep his gun ready to hand. Not wishing to be in such a situation a moment longer than he could help, he did not stop off for a few days at certain points, as he had planned, but went straight down to Barra.

He found it had been transformed in the four years he had been away. The rubber boom had begun, and the place was now the chief city of a new province. River steamers were running regularly from Pará up to it, and beyond. When you sat down to a meal, the biscuit might come from Boston, the butter from Cork, ham or codfish from Oporto, potatoes from Liverpool. The steamer service was a boon. Spruce began to plan the next leg of his botanising. This time he proposed to go up the main southern branch of the Amazon, the Solimões, on the next available steamer, bound for Peru and the Andes.

Spruce was ready for a change of scenery. The threats to his life had shocked, but not disillusioned, him. There was still so much to see and collect, from the smallest moss to the largest tree. In spite of his illness and the length of time he had been in the Amazon, he continued to maintain an astonishing work rate and work ethic. He set off upriver, marvelling at the speed of progress in the wood-fuelled steamer: 1,500 miles in eighteen days. Then it was back to canoe transport, and a wearying delay putting the boats in order and collecting a crew. But he was eventually rewarded with his first sight of the Andes, and soon he was established at Tarapoto, in north-eastern Peru, situated on a plain at about 1,500 feet, with ridges and peaks all around. Serious botanising meant a number of expeditions from this main base. On one of these, Spruce had been recommended by a local padre to an Indian named Chumbi, who agreed to put a house at his disposal. Spruce set off one morning with his temporary English assistant, Nelson, to 'herborise', while Chumbi went to shoot a bush-turkey. Chumbi's son soon came running to tell them to return;

12. The Mountains north of Tarapoto. (Spruce, August 1856)

back at the hut, they found Chumbi sitting on a log, deadly pale, and moaning with the pain of a snake-bite on his right wrist.

By this stage Spruce had acquired a good knowledge of snake-bite, and a range of possible remedies. He tried bandaging the arm; and Nelson attempted to suck out the poison. They gave Chumbi three glasses of camphorated rum, and soaked the wound in it. Then they walked him relentlessly up and down, supporting him between them. More spirits, with added quinine; strong coffee at short intervals; but still the swelling grew, till the arm was as discoloured and swollen as the branch of a tree, and the hand like a turtle's fin. The relatives were in despair; Chumbi himself chose his own burial spot, and gave directions to his wife about his children and property. Spruce, however, persisted. He made a thick poultice of rice, fomented with an infusion of herbs, and kept applying this for two days. Then, because of internal bleeding, he made a powerful diuretic from an aromatic pepper. By the third day, Chumbi was on the road to recovery, and after two months was free of symptoms. He told Spruce that it was a parrot snake, whose bite was considered incurable. Spruce was as nervous about the whole process as his friend. He could hear the whispers from the relations: *he*, Spruce, had sent Chumbi into the forest, and *he* had wished that the snake might bite him. When the padre learned of the incident, he looked very grave: 'If Chumbi had died, I should never have seen you more. Chumbi's relatives would have poisoned you,' he told Spruce. In spite of his attempts to explain the Christian doctrine of sin and death, the Indians retained a firm belief that death was in every case the work of an enemy – a view which Spruce seems to have shared. On his subsequent journey, from Tarapoto to Canelos, his men saw a small white alligator basking in the sun, and killed it with their lances:

> His stomach was distended by some food he had taken, and on piercing it, a snake's tail protruded. I laid hold on it and drew out the snake, which was closely coiled up; it was still alive (!), though so much crushed below the head as to be unable to move away. It was a terrestrial species, not venomous – yellow with black spots on the back. The body thick, passing abruptly into a short slender tail – full 3 feet long, and its destroyer no more. Thus we go on preying on each other to the end of the chapter. This poor snake, while watching for frogs among the moist stones and roots, little dreamt he was about to serve for an alligator's meal; nor the alligator, while devouring it, that he himself would soon be eaten up by Indians [and, as it happened, by Spruce].

This was an uncomfortable period in Spruce's travels. Nelson had a violent temper, and had even been arrested for murder in Peru; Spruce slept with a pistol in his hammock, and got rid of him as soon as he could, with three months' wages: he was killed on his way back to Barra.

Spruce spent eighteen months or so of solid collecting in Peru – and another three in being ill, and in preparing for his next long trip. By this time he had catalogued over 1,000 flowering plants and ferns, besides several hundred of his beloved mosses and hepaticae. He now contemplated an arduous, and dangerous, 500-mile journey to the Ecuadorean Andes: first down the tributary to the main river, then upriver by canoe towards Canelos, until the water level became too low and they had to struggle ahead on foot. He had the protection and company of two Spanish merchants for some of the way: they took turns with Spruce to mount guard at night, so that their Indian guides and porters could sleep – they had good reason to fear an attack from the Huambisas, unconverted Indians who were in conflict with the converted Indians of the little settlements. The canoes were deliberately left open, without cabins, in case they were swamped, so there was no protection from the rain; Spruce's dog Sultan, a good companion and useful guard, nearly drowned in one whirlpool, and became so deranged by the experience that he had to be shot. To begin with, there was sufficient game in the forest to keep them fed; tapir was the easiest to kill, and the most palatable. Later on, food became very scarce, and at one point they went for fifteen days without seeing a soul. At Puca-yacu, Spruce had to wait for three weeks, while enough people to carry his loads were collected: it gave him time to appreciate the view, when the weather was clear enough to see it:

> At my feet stretched the valley of the Bombonasa . . . beyond stretched the same sort of boldly undulating plain I had remarked from Andoas upwards, till reaching one long low ridge, perhaps a little higher than Puca-yacu, of remarkable equable height and direction (north to south); this is the watershed between the Bombonasa and Pastasa, and the latter river flows along its western foot; a little north of west from Puca-yacu, the course of the Pastasa is indicated by a deep gorge stretching west from behind the ridge. This gorge has on each side lofty rugged mountains (5000 to 6000 feet), spurs of the Cordillera; one of those on the right is called Abitagua, and the track from Canelos to Baños passes over its summit. [This is the route Spruce would take.] All this was frequently visible, but it was only when the mist rolled away from the plain a little after sunrise that the lofty Cordillera beyond lay in cloudless majesty.

> To the extreme left (south), at no very great distance, rose Sangahy (or the Volcan of Macas, as it is often called), remarkable for its exactly conical outline, for the snow lying on it in longish stripes, and for the cloud of smoke almost constantly hovering over it. A good way to the right is the much loftier mountain called Los Altares, its truncated summit jagged with eight peaks of nearly equal elevation and clad with an unbroken covering of snow, which glittered like crystal in the sun's rays, and made me think how pure must be the offering on 'altars' to whose height no mortal must hope to attain. Not far to the right of Los Altares, and of equal altitude, is Tunguragua, a bluff irregular peak with rounded apex capped with snow, which also descends in streaks far down its sides. To the right of Tunguragua, and over the summit of Mount Abitagua, appeared lofty blue hills, here and there painted with white; till on the extreme right was dimly visible a snowy cone of exactly the same form as Sangahy but much more distant and loftier; this was Cotopaxi, perhaps the most formidable volcano on the surface of our globe. Far behind Tunguragua, and peeping over its left shoulder, was distinctly visible, though in the far distance, a paraboloidal mass of unbroken snow; this was the summit of Chimborazo.

On clear nights Sangahy could be seen vomiting out flame every few minutes, and most days Spruce heard it exploding like a distant cannon.

The next stage of the journey was really tough. The merchants had gone on ahead, and the local governor had difficulty finding enough *cargueros* to carry Spruce's luggage – they had not unreasonably been put off by the size of his wooden boxes and leather trunks. Each *carguero* had a boy or a young woman with him, to carry his food – it was a party of sixteen. The weather was bad, with intermittent rain night and day; the track ill-defined, rough and muddy, with rapidly alternating and viciously steep ascents and descents. In spite of the slipperiness, Spruce found a pair of india-rubber shoes which he had acquired very useful; leather shoes would have been ruined, and even when these ones filled with water his feet were never cold. Once one of the streams was too full to cross, and they had to wait for the level to fall, but Spruce was reconciled to the delay by finding himself in the most mossy place he had yet seen. 'Even the topmost twigs and the very leaves were shaggy with mosses, and from the branches overhanging the river depended festoons of several feet in length, composed chiefly of Bryopterides and *Phyllogium fulgens*, in beautiful fruit.' He could forget all his troubles 'in the contemplation of a simple moss'.

In this primeval forest, between the Pastasa and the Napo, Spruce was conscious of his great predecessors: La Condamine, Bonpland and Humboldt. He was traversing the very area where Gonzalo Pizarro and Francisco de Orellana had parted in 1542; in this forest, in 1769, Madame Godin had wandered, alone and delirious, for nine days, before being found by Indians and sent forward on her solo descent of the Amazon. Even now, mid-century, he was making a journey rare enough for Wallace, his eventual editor, to comment in 1908 that he was probably the only English traveller to take that route in the past fifty years.

Spruce had serious doubts about the impact of the missionaries on the local tribes. Like Wallace – and in marked contrast to Darwin – he noted the change in the Indians' nature which seemed to go with the new religion. As soon as they had been brought to 'Christianity' they stopped making the things they needed to live on – canoes, hammocks, blowing-canes – and bought them instead from the *infideles*. Now, at a Jibaro settlement, he found that this particular group had renounced Christianity. The headman, Hueleca, was a young man of middle stature, 'slender in body, but with remarkably muscular arms and legs'. Compared with the 'Christian' Indians from Sara-yacu, Spruce found him 'a person of gentlemanly manners and with none of the craving selfishness of those people. I had therefore quite a pleasure in offering him such little presents as I had kept in store for that purpose.' All of them were complaining of illness – the children were suffering from catarrhal fever. Soon after, Spruce learned that Hueleca's wife and one of the children had died: Hueleca burned down his house and the dilapidated convent, and moved to another part of the forest 'where the whites never pass, for to their contamination he believes that he owes his bereavement'.

The last part of the journey, Spruce informed Bentham, was by far the worst:

> Road there is none, but only the merest semblance of a track, such as the tapir makes to its feeding- and drinking-places; often carried along the face of the precipices, where had it not been for projecting roots on which to lay hold, the passage would have been impossible. No one ever opens the road – no fallen trees have been cleared away – no overhanging branches cut off. From Canelos the rains set in with greater severity than ever – the dripping forest, through which I had to push my way, soaking my garments so that towards evening my arms and shoulders were quite benumbed – and the mud, which even on the tops of the hills was often over our knees – made our progress very slow and painful.

> The Indians were little accustomed to carry burdens – some of them had never been out before – and though I had made the loads as light as possible, they grumbled much and often threatened to leave me. I had brought from Tarapoto a trunk full of paper for drying my plants, but when we reached the Jibaro settlement, where unceasing rains kept us three days, I found it absolutely necessary to throw all the paper away if I did not wish to be deserted.

This was bad enough; but at the cataracts of the Topo the water was so high that one of the rocks normally used as a support for the rough bamboo bridges was completely submerged. They waited two days and nights for the level to drop; finally, out of sheer hunger, they constructed a bamboo bridge higher up, but the forty-foot span nearly broke under the weight of a man, and Spruce had to leave all his boxes behind, under a hastily made lean-to shelter. It was another three days before they reached Baños, and Spruce spent an anxious two weeks before his stuff could be retrieved: his books and journals, drugs and instruments, his Peruvian mosses and other things which no money could replace. He had been a hundred days on the voyage. He was 'much fallen in flesh', he told Bentham, and his thin face was nearly hidden by a three-months beard.

Nevertheless, though suffering from the cold – a sharp change from the Amazon basin, 48 degrees Fahrenheit at daybreak, and never more than 64 degrees, for Baños was 5,500 feet above sea level – and attacked by a violent cough which made him bleed, he was soon at work. He had no paper, but found a supply of coarse calico and began to dry the mosses and ferns from his garden walls. Some of the mosses had turned up concealed in his bedding, which did get across the river, and he started to arrange these systematically as a welcome diversion. Then his belongings arrived, and he located a supply of paper at Guayaquil and began to make plans to explore the new environment: the volcanic cone of Mount Tunguragua was on his doorstep, scarcely visited by any botanist. He remained for six months, making numerous excursions; then he moved on to Ambato, 'the prettiest town in Ecuador', which was to become his main centre for another two and a half years.

There were disadvantages. Ideas about sanitation were lax. Spruce commented that you needed to proceed with cautious steps and slow to avoid the *quisquilia* and he could never get used to people squatting at the street sides, like so many toads, and for 'a decent-looking woman in that position to look up in your face as you pass her and give you the "Buenos dias, Señor!"' Then at ten in the morning the wind would get up, and blow with the fury of a hurricane, sending a fine sand to penetrate the narrowest chink. When he rode out, he needed a gauze veil to protect the face from the combination of sun and sand and piercing wind: he had lost the skin of his nose ten times. There was a constant threat of earthquakes: Ambato had been rebuilt since the shock of 1797, and would be levelled again in 1949. But the climate was healthier than he had been used to, and he found, for the first time since he had been at Barra, congenial company. There was Dr James Taylor a day's ride away at Riobamba, the former medical attendant to President Flores, and Lecturer in Anatomy at the University of Quito – 'a very kind-hearted, honourable man, which can't be said of many Englishmen I have met in South America,' he wrote to his friend John Teasdale; and the American Minister to Ecuador, Philo White, lived at Ambato with his wife for nine months of the year. White, like many diplomats, was apt 'to run into long-winded dissertations, not remarkable for either depth or brilliancy', but he was amiable and sound-hearted, and Mrs White was a very friendly, chatty lady. Spruce would often step over to their house in the evening: 'We have, however, no chess-playing, and, instead, we rail against the people of the country – after the fashion of foreigners in all countries – and I listen patiently to Mr White's lectures on political aspects and complications.' There was, too, his landlord, Manuel Santander, and his family, who became close friends. Some years later he wrote to Spruce, recalling their friendship: 'I preserve in my heart the image of Señor Ricardo, but this my joy is troubled by the hopelessness of ever seeing him again. What happiness it would be for us to have you at Ambato just now, in the most agreeable season of the year. The time of ripe pears and peaches is near; our friend Mantilla, with his accustomed kindness is waiting for us to go and eat them. Miraflores is now planted with poplars all along the avenue where we used to walk. . . . Isobel is at the gate waiting for you. Frank and I are ready to accompany our dear friend. But – sweet dreams – delusive hopes – where is he?'

If Ambato was not quite the paradise of Señor Santander's affectionate memories, it was the most attractive location Spruce had yet found on his exhausting and debilitating travels, and he began seriously to consider a more permanent position. He was acutely conscious of his physical limitations: scrambling about on the upper slopes of the Andes, at up to 13,000 feet, exhausted him. Resting on the grass on Mount Pichincha, lying in the sun with his eyes closed, he felt what seemed like a flag waving above his face; looking up, he saw an immense condor sailing over him a few feet away. For his guide, it was a brush with death: the condor had taken the two men for corpses, and a minute later would have started to feed. Spruce was more interested in the scientific dimension. The incident, he informed Bentham, was additional confirmation that the vulture tribe hunt by sight and not by scent. Bentham had suggested he return to England, to distribute his mosses; the business of identification, sorting and sending out all the plants Spruce dispatched was obviously becoming onerous. But Spruce was afraid of falling into even more delicate health there. He had, besides, no funds beyond what Bentham had saved for him: 'these would soon be exhausted, and poverty is such a positive crime in England that to be there without either money or lucrative employment is a contingency not to be reflected on without dread.' 'I have often wished I could get some consular appointment here,' he hinted, 'were it only of £150 a year.' There was a real need for someone to watch over the interests of Europeans on the upper Amazon, and Spruce had a good knowledge of the country and the languages. If only England possessed the magnificent Amazon valley instead of India, he added, in a slightly uncharacteristic bout of imperialism: 'If that booby James, instead of putting Raleigh in prison and finally cutting off his head, had persevered in supplying him with ships, money, and men until he had formed a permanent establishment on one of the great American rivers, I have no doubt but that the whole American continent would have been at this moment in the hands of the English race.'

No consular post was offered; but the imperial connections gave Spruce his next great task. The Bentham/Hooker circle went to work: the India Office approved the appointment of Spruce to obtain seeds and young plants of cinchona – Peruvian bark – to ship out to the East, where it was planned to grow them in plantations on a large scale. The commission became Spruce's major task for the following year.

This was plant-hunting on a major scale. Quinine, derived from the bark of the red cinchona, was already highly valued as a medicine to combat fever, and was becoming recognised as crucial in the control of malaria. It was also in demand for tonic water. Supplies of bark were expensive, and getting harder to obtain: the basic method was to chop down a tree and strip off the bark, with no thought for replacement. It made sense to experiment with plantations in other countries; and Spruce was on the spot. The initiative originated in India, though Sir Joseph Banks had suggested it years before, and Sir William Hooker's elder son had written a thesis on cinchona in 1839. Dr John Royle, of the East India Company Medical Board, and Dr Thomas Anderson, superintendent of the Calcutta Botanic Garden, agreed that a collector should go to the Andes. Clements Markham, then a junior clerk at the India Office, volunteered; Sir William Hooker suggested that Spruce, already there, should be employed, and sent out Robert Cross, who was on the staff at Kew, to help him. Four areas were identified, and a budget of £500 an area assigned.

From Markham's account, one would imagine that he was the mastermind behind the scheme; but even before Markham became involved, Spruce was carrying out a preliminary reconnaissance. This entailed first a trek across the grassy plateau below Chimborazo, at a height of 12,000 feet or so, with a small string of horses and mules; then a push with a local expert into the forests of Alausí, down a track no one had used for two years, and a few nights in a temporary hut in a clearing, where Spruce lay in his hammock under the moonlight and listened to the snuffling of the bears. The only cinchona trees they found there were dead, lying on the ground stripped of bark; but eventually he did discover one with a slender twenty-foot-long shoot springing from the root. At once, he put to the test a report that he had heard, but not credited, that the trees had a milky juice: 'Berneo made a slit in the bark with the point of his cutlass, and I at once saw what was the real fact. The juice is actually colourless, but the instant it is exposed to the air it turns white, and in a few minutes red. The more rapidly this change is effected, and the deeper is the ultimate tinge assumed, the more precious is the bark presumed to be.'

1860 was the year of the cinchona expedition proper. Conditions could not have been more difficult: there were border disputes between Ecuador and Peru, and incipient civil war in Ecuador itself, with two armies and bands of deserters to contend with. At one point, the victorious government army descended on Ambato, and Spruce locked up his belongings and hung out a Union Jack, though the danger passed on this occasion: the army pillaged the neighbouring town instead. He spent several months in protracted negotiations. The authorities were extremely reluctant to grant rights, but he enlisted his friend Dr James Taylor, and this helped persuade one of the landowners, ex-President Flores, who conveniently had been Taylor's patient. For $400, Spruce would be allowed to take seeds and plants from the forest, but no bark, and he would be given the assistance of four Indian helpers. Flores presumably knew the potential value of the cinchona, though, according to Spruce, most of the Indian population thought it was used for dye. When Spruce told them that the bark yielded the quinine which was so useful in medicine, he heard them say afterwards: 'It is all very fine for him to stuff us with such a tale; of course *he* won't tell us how the dye is made, or we should use it ourselves for our ponchos and bayetas, and not let foreigners take away so much of it.'

Spruce's health was bad in the spring of 1860. He became temporarily deaf – perhaps from taking too much quinine – and on 29 April he woke up paralysed from rheumatism in his back and legs. But with Taylor's encouragement – he told Spruce the warm climate of the forest would be good for him – they set off in June for the Red Bark forests of Chimborazo, on the western slopes of the Andes. They made their base near Limón, in a timber and bamboo cane-mill, 'abundantly ventilated', as Spruce commented, 'and only too frequently filled with fog, as we found to our cost, in coughs and aching limbs, and in mouldy garments, saddles etc.' Spruce settled down to watch for the critical moment when the fruit capsules would ripen, while Taylor rode down to Ventanas to meet Robert Cross. Cross was travelling with the materials for a set of fifteen Wardian cases in which the seedlings were to be transported. Spruce, watching over his precious crop, received an unpleasant shock: checking the trees one morning, he found two completely stripped of the as yet unripe seed capsules. The locals had heard that Spruce would buy seeds, and had begun harvesting in anticipation. He had to explain that only seeds collected by him in person would have any value, and arranged payment to the local landowners to guard the trees. While he waited patiently for the seeds to ripen, he made a comprehensive study of the rest of the forest vegetation.

Cross, who had been delayed first by illness and then by the civil war, arrived at Limón at the end of July. They set to work to fence in some land, and made a pit to hold over a thousand cuttings. These were carefully protected and nurtured – they used calico as a substitute for glass – and when the sun emerged fiercely from the mist, the seedlings needed to be kept damp, which sometimes entailed a frantic chain of buckets. But Cross was obviously an expert and meticulous nurseryman, and Spruce was able to leave him in charge while he went to gather seeds from another location: together, they collected about 100,000. Spruce took one batch down to Guayaquil, for shipment to Jamaica, then rode back to help Cross with the next stage. The plants had to be packed out in panniers, strapped to oxen, and taken to the riverside. There the fifteen Wardian cases were assembled, the plants – 637 by this time – were installed, and everything loaded on to a raft. Luckily, they decided not to put the glass in at this stage, for the raft suffered at least one major collision on its swift passage downriver. Finally, on 31 December, the cases were loaded aboard, and Spruce was happy to see them go in Cross's care, bound first for Lima and Panama, and then to their final destinations in India and Ceylon. Four hundred and sixty-three seedlings survived, to form the nucleus of an experimental plantation near Ootacamund in the Nilgiri Hills. A few months later, a law was passed in Ecuador banning the export of plants.

This might have been the moment for Spruce to return home, but he was destined for an enforced delay. He had entrusted his savings, some £700, to a local firm, which collapsed. Partly as a consequence, he spent three more rather frustrating years collecting in Peru.

His achievements were immense. He had collected more than 30,000 specimens, explored areas of the Amazon and Andes for the first time, and recorded twenty-one Indian vocabularies. He had, besides his success with the cinchona, sent back vital botanical information about the *hevea* rubber plant; his knowledge of the medicinal properties of a number of plants was exceptional – for example, a sample of coca he had sent to Dr Nieman in Berlin led to the isolation of the active principle of alkaloid cocaine in 1858, and the material he had dispatched to Kew was unsurpassed. He was honoured by a doctorate from the Imperial German Academy in 1864: he might, had he not been so modest, have expected some recognition in England. It took a campaign by Wallace, with the help of Darwin and Hooker, to secure him a pension of £100 a year.

Spruce lived quietly in Yorkshire at Welburn, and then Coneysthorpe, small villages on the Castle Howard estate. He was so crippled in health that he found it difficult to sit at his desk for any length of time. But he went on working, corresponding, writing a few papers, entertaining his friends, and producing a monumental study of his speciality: *The Hepaticae of the Amazon and the Andes of Peru and Ecuador*, which included 700 species, 400 of which had been first named and described by himself. The plaque on his cottage reads:

Richard Spruce
1817–1893
of Ganthorpe, Welburn
and Coneysthorpe
Distinguished botanist, fearless
explorer, humble man, lived here
1876–1893

13. Richard Spruce, aged seventy-two.

He never, characteristically, published an account of his travels; after his death his old friend Alfred Wallace took on the task, and *Notes of a Botanist on the Amazon and Andes* finally appeared in 1908. (Wallace even called in a clairvoyant in an attempt to track down some of the missing journals; 'Cannot you get a strong woman to make a *thorough search* from *attic* to *cellar*,' he wrote to Spruce's executor, 'in *cupboards, boxes & bundles, & everywhere else*, till they are found.')

Of the three men who had dined together at Captain Hislop's table in 1849, all great naturalists, Richard Spruce had perhaps the purest scientific curiosity. He loved living things for their own sake, and understood most acutely the delicate balance of life in the rain forest. When he had to leave for Panuré from the highest point he had reached on the upper Uaupés, on the border between Brazil and Colombia, the weather suddenly cleared, and, as he wrote regretfully to George Bentham, 'I well recollect how the banks of the river had become clad with flowers, as it were by some sudden magic, and how I said to myself, as I scanned the lofty trees with wistful and disappointed eyes, "there goes a new *Dipteryx* – there goes a new *Qualea* – there goes a new the Lord knows what!" until I could no longer bear the sight and covering up my face with my hands, I resigned myself to the sorrowful reflection that I must leave all these fine things "to waste their sweetness on the desert air".'

5 *The Plant-hunters*

THE WORLDWIDE TRADE IN PLANTS expanded enormously in the nineteenth century. Europeans regarded the rest of the world, or at least its most fertile areas, as an extended farm, and were constantly looking for useful and financially viable species to transplant: tea from China to India, rubber from Brazil to South-east Asia. Captain Bligh on the *Bounty* was dispatched at Sir Joseph Banks's suggestion to collect breadfruit seedlings from Tahiti, and ship them to the West Indies as a cheap source of food. Five months spent waiting for the young plants to grow sufficiently robust to survive the voyage proved too powerful a lure for the crew; when they mutinied, one of their first actions was to drop the breadfruit seedlings over the side. The wastage on all transported plants, and seeds, was staggering; added to the dangers and difficulties which plant-collectors had to endure at source, especially in China and Japan, was the damage caused by violent changes in temperature and humidity as well as the effects of salt water and the depredations of rats. Fortunately the aristocratic patrons who helped to fund the network of collectors, so as to stock their private orangeries and glasshouses, had very deep pockets.

The great breakthrough in plant transportation came in 1834, with publicity about the invention of the Wardian case. Dr Nathaniel Ward was a Whitechapel surgeon. He noticed that out of a little moist mould left some months back, together with a pupa of a hawkmoth, in a sealed glass jar, a grass seedling and the sporeling of a fern had begun to sprout. He calculated that this might be a way to nurture plants in the poisonous air of central London: sealed up with moisture which they would recycle, they would be better protected, and would not require constant attention. Loddiges the nurserymen made up a number of glass jars, and Ward's indoor garden flourished. The landscape expert and horticultural writer Loudon went to admire it. He thought it 'the most extraordinary city garden' he had ever seen, and enthused in the *Gardener's Magazine* about the extensive application of the principle 'in transporting plants from one country to another; in

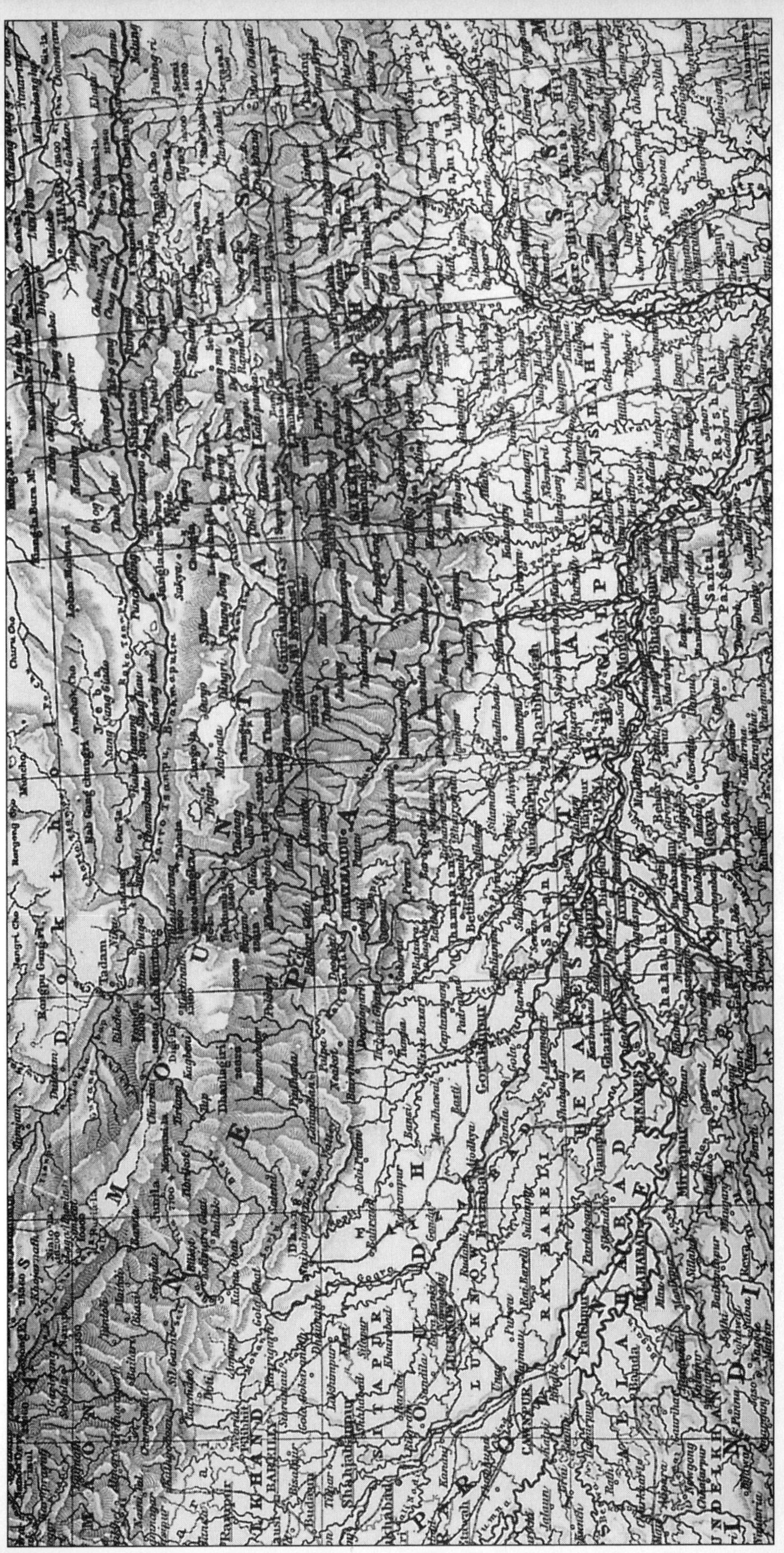

Nepal. Sikkim and Bhutan (The Times Atlas, 1895)

preserving plants in rooms or in towns; and in forming miniature gardens or conservatories'. The glass container made plant-growing a much more accessible pastime: you no longer needed to be the Duke of Devonshire or even a well-heeled country parson with a conservatory to cultivate exotic plants. The Wardian case led to a serious outbreak of fern cultivation among the middle classes; and Ward extended the principle to the aquarium, by introducing a fishtank into his fern-house and demonstrating that the fish could survive in conjunction with aquatic plants. He added a chameleon and a Jersey toad, which lived for ten years.

The Wardian case revolutionised the shipment of plants, and gave a great spur to the search for increasingly rare and delicate flowers, such as orchids, to grace the glasshouses, conservatories, stoves and winter-gardens of Europe. In 1826 the Duke of Devonshire appointed Joseph Paxton to be his gardener at Chatsworth, and one of the most productive of gardening partnerships was born. These men thought on the grand scale, as the Great Conservatory at Chatsworth, Paxton's masterpiece in conjunction with Decimus Burton, and the forerunner of the Crystal Palace, would demonstrate: completed by 1841, it was 277 feet long, and heated by seven miles of pipes, with coal brought to feed the eight boilers by underground railway. But just as important as these huge tropical worlds were the plants to stock them. In 1835 Paxton selected John Gibson, a young gardener on his staff, to go out to the Khasia mountains in Assam on an orchid hunt. As a quid pro quo, Gibson would escort out a collection of plants and seeds for the Calcutta Botanic Garden, the plants duly installed in the new airtight cases. The Duke arranged for Gibson to travel out on HMS *Jupiter*, which was carrying Lord Auckland towards his appointment as Governor-General, and a courteous note to Dr Wallich, the superintendent of the Calcutta Garden, ensured that Gibson would be given every assistance.

Gibson's arrival in Calcutta coincided with the flowering of the holy grail of Asia specimens, the *Amherstia nobilis*. When Gibson saw this magnificent sight, he apparently ran round it clapping his hands 'like a boy who has got three runs in a cricket match'. While Gibson was packed off to the hills to collect the orchid harvest, Wallich planned a further trip for him to procure plants and possibly seeds of the *Amherstia* at Martaban, in Burma. But Gibson was ecstatic up at Chirra Poonje. 'I never saw, nor could I believe that there was such a fertile place under the Heavens.' Almost every plant was new to the European collections. He sent load after load back to Calcutta, and reckoned that he had found between eighty and ninety new species of orchid. Before he left in March 1837, he had supervised the packing of two plants of *Amherstia* from the Botanic Garden, one for Chatsworth, and one for the East India Company – a much more convenient source than a trip to Burma. 'One I put in the chest myself yesterday,' he wrote to Paxton, 'the other has been packed some time. I must say that no care or attention whatever shall be wanting on my part to keep these, the most noble of all plants alive. Nothing can be more beautiful on earth than the plant in question.'

The orchids all arrived in good shape; but the Chatsworth-bound *Amherstia* died. The Duke applied to the directors of the East India Company, and they gracefully yielded their own healthy specimen to him. Paxton set off for London and Devonshire House. The Duke ordered his breakfast to be served in the Painted Hall, and ate it while he introduced Paxton to the *Amherstia*; Paxton was followed by a series of other plant-lovers, such as George Loddiges and Dr John Lindley, the assistant secretary of the Horticultural Society and Professor of Botany at London University. Then Paxton and Gibson travelled north to Chatsworth with their treasure, still planted in its heavy case, and installed it in the specially built stovehouse. There it flourished, but never flowered, one wonder among the astonishing display of tropical plants living inside their glass palaces in the heart of the Derbyshire countryside. In 1843, Queen Victoria and Prince Albert drove in an open carriage down the Great Conservatory's central aisle; 12,000 lamps blazed out in the December night. In the bespoke lily-house, the Queen was able to admire the giant water-lily named after her, *Victoria regia*. Inspired by the richness of the Indian forests, Gibson later recreated something of their splendour for Londoners in Battersea Park, where, as superintendent in 1854, he oversaw a layout which included a fifteen-acre lake, alpine and rock sections, and a subtropical garden.

As a plant-hunter, Joseph Hooker was supreme. He became the outstanding English geo-botanist of the century, and his distinction was founded upon an unrivalled basis of experience on the ground. After accompanying Ross to the Antarctic and so acquiring a first-hand knowledge of the flora of the southern hemisphere, he was already at work describing the Niger Flora and was eager to see some area of the tropics for himself. This burning ambition even took priority over his love for Frances Henslow, a daughter of the Cambridge professor.

Hooker badgered Ross, and investigated the possibility of a naval expedition to Borneo and one to Goa courtesy of the East India Company. The final choice lay between the Andes and the

Himalayas. Hugh Falconer, the newly appointed superintendent of the Botanic Garden at Calcutta, persuaded him that the Himalaya range was the place to go – that 'we' (the British) 'were ignorant even of the geography of the central and eastern parts of these mountains, while all to the north was involved in a mystery equally attractive to the traveller and the naturalist'. Armed with a letter from Humboldt, and the support of the Chief Commissioner of Woods and Forests, Hooker was promised £400 per annum for two years; he had one preliminary task, to investigate fossil plants for the Geological Survey, and a major commission to collect plants for Kew. Thus financed, he was headed for Sikkim, and the approaches to Tibet. The third year was to be spent in Borneo, investigating the suitability of Labuan for growing crops such as cotton, tobacco, sugar and indigo.

Hooker travelled out in November with some unusual intellectual property stored in his mind. Earlier in the year Darwin had given him a copy of his evolution essay. Hooker was by this time Darwin's trusted sounding-board, someone the older man could share his ideas with, and someone from whom he would accept questions and criticisms of detail. Darwin was as reluctant as Frances Henslow must have been to see Hooker leave, and bombarded him with letters and commissions during his tour. The great speculations about landforms, about varieties and species and their distribution, can be detected in Hooker's letters and in his *Himalayan Journals*, a complex and recurrent theme which surfaces from time to time from underneath the day-to-day pattern of observation. Back at Down, Darwin immersed himself in detailed work on a single species, the barnacle.

Hooker may not have had much family money behind him, but he certainly had influential friends. He travelled to Egypt on HMS *Sidon*, whose main passenger was the Marquis of Dalhousie, *en route* to take up his post as Governor-General of India. By the time they transferred to the Indian navy's steam frigate *Moozuffer*, Hooker had been invited to consider himself a member of the Governor's suite (though even in that privileged status he had to pig it in the ship's armoury, 'next to the engine, intolerably hot and smothered with coal-dust'). When he reached Calcutta he was bidden to make Government House his Indian home. As he commented, this connection ensured him 'the best reception everywhere'. Lord Dalhousie he found extremely agreeable and intelligent in everything except natural history and science – a 'perfect specimen of the miserable system of education pursued at Oxford' – but without his support, he would never have reached his Himalayan goal. He was shown round the Botanic Garden, among whose triumphs was the successful introduction of the tea-plant from China. He was presented with a fully equipped *palkee* (a more elaborate Indian version of a sedan chair) by Sir James Colville, president of the Asiatic Society. The full political safety net was securely in place before he set off on the first section of his journey, a preliminary swing through Bengal which took in an investigation of the coal-beds of the Damooda valley.

The first impression of Hooker which emerges from his *Himalayan Journals* is of a rather conventional Englishman, as he is carried by Indian bearers down the grand trunk road. The novelty, and the first illusion of comfort, rapidly wears off. Instead, we hear slightly testy colonial complaints about 'the greedy cry and gestures of the bearers, when, on changing, they break a fitful sleep by poking a torch in your face, and vociferating "Bucksheesh, Sahib"; their discontent at the most liberal largesse, and the sluggishness of the next set who want bribes, put the traveller out of patience with the natives'. The disillusion with India, or with Indians, is rapid, and there is a good deal more comment of the 'natives are far from honest' variety. But at this point Hooker is accompanying official representatives, surveyors, magistrates and revenue collectors; he is trampling through the hills with a small army of elephants and bullock-carts, pushing through the jungle with loads of mining equipment; he takes part in a tiger shoot, without success; he recounts lurid stories about Thugs, dacoits and poisoners.

At Mirzapore, he arrived at the Ganges and changed his transport to a boat, going downstream to Benares and Patna, where he examined the opium godowns. His description makes unnerving reading. The British certainly pioneered the large-scale manufacture of opium, applying to it the full range of Victorian factory techniques: the workmen were hosed down at the end of the day's labour, and the 'inspissated' water was used as liquor to mould the next day's output of opium balls, so that not a particle was lost. A powerful smell pervaded the buildings, but, he was assured, this did not affect Dr Corbett, the assistant-agent, or his assistants. A workman could produce anything between thirty and fifty balls in a ten-hour day; the total produce was over 10,000 a day, and each season 1,353 million balls were manufactured for the Chinese market alone. Hooker was clearly impressed by the supreme efficiency of this process, with both growers and dealers strictly licensed; his only comment about the end-user is to suggest that by the time it reached the retailers in the bazaars, the opium was so adulterated that it only retained one-thirtieth of its intoxicating power in its pure state.

Having visited one of the commercial treasure houses of the

subcontinent, Hooker moved on to Bhagalpore, where he enthused about the Horticultural Gardens. He concluded with a wonderfully establishment commendation: 'Such gardens diffuse a taste for the most healthful employments, and offer an elegant resource for the many unoccupied hours which the Englishman in India finds upon his hands.'

As he began to move north, and higher, towards the Himalayas, Hooker's spirits rose, and his narrative takes on a more buoyant tone. The forest along the steep mountainsides was 'truly magnificent'. Nature was operating on a gigantic scale. 'Vapours, raised from an ocean whose nearest shore is more than 400 miles distant, are safely transported without the loss of one drop of water, to support the rank luxuriance of this far distant region.' The soil and bushes swarmed with large and troublesome ants, and enormous earthworms; in the evening, the noise of the great *Cicadae* in the trees was almost deafening. The leeches could puncture through thick worsted stockings, and even trousers, and when full of blood rolled into the bottom of his shoes, such soft balls he hardly felt them in walking. The complaints of the hot plains vanish, and he makes light of any discomfort.

A pony was sent from Darjeeling, and on it Hooker climbed steeply towards Pacheem.

> The prevalent timber is gigantic, and scaled by climbing *Leguminosae*, which sometimes sheath the trunks, or span the forest with huge cables, binding tree to tree. Their trunks are also clothed with parasitical Orchids, climbing Pothos, Peppers, Vines, Convolvulus, and Bigonias. The beauty of the drapery of the Pothos-leaves is pre-eminent, whether for the graceful folds the foliage assumes, or for the liveliness of its colour. Of the more conspicuous smaller trees, the wild banana is the most abundant, its crown of very beautiful foliage contrasting with the smaller-leaved plants amongst which it nestles; next comes a screw-pine (*Pandanus*) with a straight stem and a tuft of leaves, each eight or ten feet long, waving on all sides. Bamboo abounds everywhere: its dense tufts of culms, 100 feet and upwards high, are as thick at the base as a man's thigh. Twenty or thirty species of ferns (including a tree-fern) were luxuriant and handsome; while folicaceous lichens and a few mosses appeared at 2000 feet. Such is the vegetation of the roads through the tropical forests of the Outer Himalaya.

He went zigzagging up through the dripping forest to Pacheem bungalow, the most sinister-looking rest-house stuck on a little cleared spur of mountain: 'surrounded by dark forests, overhanging a profound valley, enveloped in mists and rain, and hideous in architecture, being a miserable attempt to unite the Swiss cottage with the suburban gothic'. Cold, sluggish beetles clung to the damp walls of the ghostly room, and he felt himself a benighted traveller braving the horrors of the Harz forest. But next morning he collected about sixty species of ferns, and marvelled at the astonishing profusion of forest trees: oaks, magnolias, laurels, birch, alder, maple, holly, bird-cherry, common cherry and apple. There was, too, a 'most noble white rhododendron, whose enormous and delicious lemon-scented blossoms' strewed the ground. Exhilarated, he rode on to Darjeeling, where he was based from April 1848 until October, collecting, making short expeditions, and waiting for the necessary permission for his projected trip into Sikkim.

Darjeeling, 'the Queen of Hill Stations', is situated at nearly 7,000 feet, giving an unsurpassed view of the Himalayas, with Kinchinjunga, 28,150 feet high, as the great centrepiece – then thought to be the highest mountain in the world. The vegetation was superb, and the only possible drawback was the 'Greenock-like' climate. Hooker at once found himself at home, welcomed to the house of Brian Hodgson, formerly the British Resident at the Nepalese Court. Hodgson was a zoologist, oriental linguist, ethnologist and geographer, and proved a good friend and mentor for Hooker, who was soon installed in his hilltop bungalow with its broad veranda facing north towards the 'Snowy Mountains'. Hooker had two rooms to himself; there were lots of servants 'to go and come as I please to call or send', innumerable cats, and more 'Bishop Barnabees' – ladybirds – than at Kew. Every evening Hooker and Hodgson worked together at Himalayan and Tibetan geography and natural history – 'though I say it myself,' Hooker wrote to his mother, 'I ought in a month or two to have a better knowledge of these aspects of India than any man, having every advantage that an excellent library and tutor can afford'. The two lived like hermits, seeing few visitors besides Campbell and his wife. Dr Campbell was the political agent to Sikkim, and he too became a close friend, while Hooker grew devoted to the beautiful Campbell children, and was godfather to the Campbells' next child, Josephine. Hooker marked his gratitude by naming new species after his friends: a magnolia after Hodgson, a great white rhododendron after Lady Dalhousie, another rhododendron to be called MacCallum Morae for Mrs Campbell. 'Now pray,' he asked his father, 'don't forget to attach the name to one of the species sent if the one I have given it be not new.'

These months offered invaluable training for the more arduous time ahead. Hooker climbed to 10,000 feet, discovered how to manage in a tent and how to cross rivers by cane bridges, and learned, crucially, how to conduct himself 'beyond British ground', in his dealings with the Lepchas who accompanied him, or with the lamas in the temples. He trained some of the Lepcha 'lads' as plant-collectors and plant-driers, and swiftly began to appreciate their qualities:

> It is always interesting to roam with an aboriginal, and especially a mountain people, through their thinly inhabited valleys, over their grand mountains, and to dwell alone with them in their gloomy and forbidding forests. . . . A more interesting and attractive companion than the Lepcha I never lived with: cheerful, kind, and patient with a master to whom he is attached; rude but not savage, ignorant, and yet intelligent; with the simple resource of a plain knife he makes his house and furnishes yours, with a speed, alacrity, and ingenuity that while away that well-known long hour when the weary pilgrim frets for his couch. In all my dealing with these people, they proved scrupulously honest. Except for drunkenness and carelessness, I never had to complain of any of the merry troop; some of whom, bareheaded and bare-legged, possessing little or nothing save a cotton garment and a long knife, followed me for many months, from the scorching plains to the everlasting snows.

At one point he was employing eighteen collectors at Darjeeling, paying them between eight and sixteen shillings a month; two rooms in Hodgson's bungalow were full of his plants, arranged on a circle of chairs around the fender, while two Lepchas squatted by the fire drying papers all day long.

What sort of pilgrimage was Hooker contemplating? Sikkim, a semi-autonomous state, was under a loose kind of British protection (protection, in effect, from the Nepalese); the Rajah had ceded a chunk of mountain land, including Darjeeling, for £300 a year in return for this privilege, and the British had installed Dr Campbell as superintendent of the station, to be responsible for political relations between the British and Sikkim governments. In recent years, the Rajah's influence had waned as that of his brother-in-law Namgay, the Dewan, increased, and political tension grew in proportion. Hooker's description bristles with indignation: 'Every obstacle was thrown by him [the Dewan] in the way of a good understanding between Sikkim and the British government. British subjects were rigorously excluded from Sikkim; every liberal offer for free trade and intercourse was rejected, generally with insolence; merchandise was taxed, and notorious offenders, refugees from the British territories, were harboured; despatches were detained; and the Vakeels, or Rajah's representatives, were chosen for their overbearing manners and incapacity.' If only, Hooker lamented, the Calcutta government had not been so pusillanimous; if only insolence had been noticed at the first sign, and nipped in the bud, and so on. The Governor-General attempted to arrange Hooker's journey. After awkward negotiations, it was agreed that he could pass through the eastern region of Nepal, west of Kinchinjunga, and then return to Darjeeling through Sikkim, after 'visiting' the Tibetan passes on the way. Hooker's objectives were, partly, to add to his collections; but they were also to map and measure; and his persistence in pushing up remote passes, and insistence on visiting alternative routes, suggests that he had been enrolled as a willing participant in the Great Game.

On 27 October, Hooker set off for his three-month expedition. The Victorian scientific explorer may have felt isolated, but he was not alone. This was a formidable logistical enterprise, because everything, including food for the porters, had to be carried on men's backs. Seven were assigned to Hooker's tent, instruments and personal equipment; seven more to the papers for drying plants. The interpreter, the headman and the chief plant collector had a man each. There was a bird and animal shooter, collector and stuffer, with four men to lug the ammunition; three Lepcha lads, whom Hooker had been training, to climb trees and change the plant papers; fourteen Bhutan porters, laden with food; Hooker's personal servant; a detachment of eight Nepalese guards – 'immense fellows, stout and brawny', in loose scarlet jackets with a kukri in their cummerbund and heavy iron sword at their side – and a nattily dressed havildar in command, plus two porters for their food. Hodgson pressed personal stores on him; Mrs Campbell rummaged through her larder and store-room, and added veils as glare protection plus a selection of woolly stockings; the Mullers, on leave from the Patna opium operation, checked Hooker's time-keepers and overhauled his instruments. The heavily laden party which Dr Campbell saw off on its route along the Goong ridge towards Nepal was fifty-six strong.

Hooker himself carried a small barometer, a large knife and digger for plants, telescope and compass, and notebook and pencil tied to his jacket pocket for instantaneous observations. Two or three of the Lepcha 'lads' stayed close to him during the trek, with botanising box, thermometers, sextant and artificial horizon, measuring-tape, azimuth compass and stand, geological hammer, bottles and boxes for insects, and sketchbook. The normal

14. The Tambur river at the lower limit of firs. (Hooker, *Himalayan Journals*)

Himalayan practice was to strike camp very early, and arrive at the next site around noon. Hooker found this unsatisfactory; he set off later, having done some collecting and measuring, then paused along the route for further collecting, before spending the evening writing up his notes and journals, and ticketing the plants gathered during the day. At first he kept the scientific observations to a minimum, using the barometer only in the privacy of his tent to allay suspicion; but whenever possible he recorded temperature, air and ground, altitude, humidity and weather conditions, in addition to his map-making measurements.

As the routine became established, Hooker eased himself into a new way of life. He found the Bhutanese the least satisfactory of his retinue; they were constantly complaining, and some of them made off and had to be replaced. The Gurkhas were 'sprightly', telling interminable stories and singing Hindu songs; 'being neater and better dressed, and having a servant to cook their food, they seemed quite the gentlemen of the party'. But still, for Hooker, it was the Lepcha who was the most attractive, the least restrained, the most natural: 'the simplest in his wants and appliances, with a bamboo as his water-jug, an earthen pot as his kettle, and all manner of herbs collected during the day's march to flavour his food'. Hooker's tent was a blanket thrown over a tree limb, to which others were attached over a rough frame. One half was taken up by his bed, with box of clothes beneath; books and writing materials were stowed under a rough table, constructed, like the bed, from bamboo by the Lepchas with their knives. The barometer was hung in the furthest corner, surrounded by the other instruments. 'A small candle was burning in a glass shade, to keep the light from draught and insects, and I had the comfort of seeing the knife, fork, and spoon laid on a white napkin, as I entered my snug little house, and flung myself on the elastic couch to ruminate on the proceedings of the day, and speculate on those of the morrow, while waiting for my meal, which usually consisted of stewed meat and rice, with biscuits and tea.' Hooker was beginning to enjoy the simple life. He was physically very tough, and positively relished the challenges and privations which the expedition gradually imposed.

Hooker was pushing up the Tambur valley to the village of Wallanchoon, inhabited largely by Tibetans. At 8,000 feet, there were firs, and many subalpine plants. The scenery was 'as grand as any pictured by Salvator Rosa; a river roaring in sheets of foam, sombre woods, crags of gneiss, and tier upon tier of lofty mountains flanked and crested with groves of black firs, terminating in snow-sprinkled rocky peaks'. Hooker was startled to emerge from

15. Hooker's drawing of Hooker drawing in Wallanchoon village.

the gorge to an open flat on the riverbank, a village of painted wooden houses ornamented by flags, and swarms of good-natured, 'intolerably dirty' Tibetans, who were very civil 'and only offensive in smell'. (This was one aspect Hooker never became used to – he had an aversion to fleas – though he obviously warmed to the people and to their way of life in every aspect except cleanliness.)

At Wallanchoon, Hooker came up against his first major obstacle. He was headed for the pass into Tibet, and carried authorisation to that effect from the Rajah of Nepal. The Nepalese officer was suddenly reluctant. The local headman denied that the Rajah had any right to grant permission. Essentials – a guide, snowboots, extra blankets, provisions – were mysteriously unobtainable. Everyone, it seems, had assumed that Hooker would have had quite enough of the altitude and unpleasant weather long before he reached this critical point. With British obstinacy, he announced that he intended to move on, with or without permission, and informed both the loaned Nepalese soldiers and his servants that their pay or reward depended on their compliance. After some bullying, further persuasion, and a little medical aid – 'the prescribing of pills, prayers, and charms in the shape of warm water' – a guide and snowboots appeared, and he set off with a reduced party. The landscape reminded him of the Scottish Highlands, especially when he found on the narrow path a common British grass, *Poa annua*, and the familiar shepherd's purse: 'They had evidently been imported by man and yaks, and as they do not occur in India, I could not but regard these little wanderers from the north with the deepest interest.' At 13,000 feet, with the ground hard and frozen, Hooker admitted to the first symptoms of lassitude, headache and giddiness – but these were slight, and only came on with severe exertion. They found a stone hut for the night and built a juniper fire: the Gurkhas retired to one corner, the Lepchas to another, while one end was screened off for Hooker's couch; unluckily, it faced north-east, and the wind poured through the cracks, making sleep impossible. The next morning they struck up towards the pass for ten miles, and then for a further four over snow. After five and a half hours, Hooker reached the summit 'utterly knocked up': the boundary between Nepal and Tibet was a low saddle between two rugged ridges, marked by a cairn decorated with bits of stick and rag covered with Tibetan inscriptions. He calculated the height as 16,748 feet by comparative observation with Darjeeling, and collected some plants, including *Saussurea gossypina*, 'which forms great clubs of the softest white wool, six inches to a foot high, its flowers and leaves seeming clothed with the warmest fur that nature can devise'. It was 18 degrees Fahrenheit; the sun was hidden behind rocks, and the wind was bitterly cold. He had arranged an intermediate camp, and they sheltered under some huge boulders in an ancient moraine; apart from an excruciating headache, he felt no ill-effects, and after a supper of tea and biscuit slept soundly.

Hooker was growing in confidence. He decided to send the majority of his party back down the Tambur valley to Darjeeling, with his precious plant collections. Meanwhile, with a reduced group of nineteen, he set out to tackle a formidable series of valleys and passes, ending up at Jongri, in Sikkim, on the south flank of Kinchinjunga. The fewer people he had with him, the more remote the situation, the more he seemed to thrive. His reserve stock of provisions was dwindling, and he was only momentarily thrown by discovering that of four tins which he thought were meat, three contained prunes, and one *dindon aux truffes*; however, the 'greasy French viand' could be used as a sauce for his staple diet of rice. On this part of the journey, Hooker seems more intent on taking angles and observations and making sketches than plant collecting. He went up the Kanglachem pass as far as he could, till the way was blocked by a huge moraine, then crossed the Yangma river and moved on to the Kambachen, and after that the Choonjerma pass. Hooker often walked ahead of his team, and relates his experience of the sublime in a passage which uses the standard European reference points of science, industry and art, and yet acknowledges their insufficiency:

> Evening overtook us while still near the last ascent. As the sun declined, the snow at our feet reflected the most delicate peach-bloom hue; and looking west from the top of the pass, the scenery was gorgeous beyond description, for the sun was just plunging into a sea of mist, in a blaze of the ruddiest coppery hue. As it sank, the Nepal peaks to the right assumed more definite, darker, and gigantic forms, and floods of light shot across the misty ocean, bathing the landscape in the most wonderful and indescribable changing tints. While the luminary was vanishing, the whole horizon glowed like copper from a smelting furnace, and when it had disappeared, the little inequalities of the ragged edges of the mist were lighted up and shone like a row of volcanoes in the distance. I have never before or since seen anything, which for sublimity, beauty, and marvellous effects, could compare with what I gazed on that evening from Choonjerma pass. In some of Turner's pictures I have recognized similar effects, caught and fixed by a marvellous effort of genius; such are the fleeting hues over the ice, in his 'Whalers,' and the ruddy fire in his 'Wind, Steam, and Rain,' which one almost fears to touch. Dissolving views give some idea of the magic creation and dispersion of the colours, but any combination of science and art can no more recall the scene, than it can the feelings of awe that crept over me, during the hour I spent in solitude amongst these stupendous mountains.

However, reality soon broke in. Such reveries were dangerous, with feet and legs wet through and the temperature at freezing point. The porters arrived, grunting and panting, and were revived by half a bottle of brandy.

The passes Hooker had wished to penetrate were closed, so he moved south, then east across the Islumbo pass into Sikkim to the Kulhait river. There a message from Dr Campbell reached him, asking him to cross to the Teesta river, and join him at Bhomsong, 'where no European had ever yet been', for a meeting with the Rajah. This proved a protracted and unsatisfactory affair, with the Dewan interposing himself, and a sequence of supposed breaches of etiquette which were interpreted as insults by Hooker and Campbell. The exchanges ended in a scoreless draw; and Campbell returned to Darjeeling, followed at a short interval by Hooker, after a scramble on the south flank of Kinchinjunga, and a visit to the Changachelling temples. The monks were painting the vestibule, and among the figures in the scene Hooker was struck by that of an Englishman, whom he soon recognised, to the artist-monk's delight, as himself: 'I was depicted in a flowered silk coat instead of a tartan shooting jacket, my shoes were turned up at the toes, and I had on spectacles and a tartar cap, and was writing notes in a book. On one side a snake-king was politely handing me fruit, and on the other a horrible demon was writhing.' On 19 January he arrived back in Darjeeling, to the relief and rejoicing of his Lepcha followers and their families.

The fruits of this first journey were dispatched to Calcutta – eighty loads. A visitor to Darjeeling, William Tayler the Post-master-General of Bengal, decided to paint Hooker. He posed him sitting inspecting a vasculum full of plants, with his Lepcha headman kneeling before him, holding a splendid bunch of *Dendromium nobile* in his hand; the Gurkha havildar is standing in the background; Hooker's Bhote mastiff, Kinchin, which he had brought back from the snows, stands on guard, while to the left two Lepchas put up the blanket tent-house and cut bamboos: 'My dress was the puzzle, but it was finally agreed I should be as I was when in my best, a Thibetan in the main, with just so much of English peeping out as should proclaim me no Bhotea, and as much of the latter as should vouchsafe my being a person of rank in the character.' This concept translated into Bhotean cloak and boots, English pantaloons, shirt collar 'romantically loose and open, with a blue neckerchief', and a pale grey felt Tibetan cap, topped with a silver-mounted pebble and a peacock's feather floating down his back, as marks of rank. The painting was later copied and tidied up in London. Out went the dog (which had, in fact, by

16. The Botanist in Sikkim. (From William Tayler's picture)

that time drowned in a river crossing) and the viewer's eye is led towards a more dominant seated figure of an unmistakably serious English scientist, receiving tribute from a kneeling suppliant: a new lord of creation, Adam in the act of naming.

After an interlude in the plains, Hooker was ready to march north again, aiming for the upper Teesta valley and the Sikkim-Tibet frontier north-east of Kinchinjunga. He made his way up the Teesta river, encountering various obstructions, although a friendly lama whom he knew from his previous trip told him to ignore everything, and reassured him that that these were the Dewan's doing, not the Rajah's. Higher up, the Teesta is known as the Lachen-Lachoong, and at Choongtam it splits into the two main tributaries. Here opposition increased; the locals were under clear instructions not to sell Hooker any provisions. Eventually, fresh supplies from Darjeeling arrived, not before Hooker had been sorely tempted to help himself from the village. He had a choice of routes to Tibet, each about six days' march distant, and decided to tackle the Lachen valley first, which led to the Kongra Lama pass, setting off for Lamteng on 25 May. Lamteng proved an interesting area botanically. The vegetation 'is European and North American; that is to say, it unites the boreal and temperate floras of the east and west hemispheres; presenting also a few features peculiar to Asia. This is a subject of very great importance in physical geography; as a country combining the botanical characters of several others, [it] affords materials for tracing the direction in which genera and species have migrated, the causes that favour their migrations, and the laws that determine the types or forms of one region, which represent those of another.' Darwin's paper on evolution was still running in his head.

In addition to the relentless routine of taking scientific observations, and collecting, and writing up his notes and journal, Hooker also kept up a lively correspondence with family and friends. The weather, he told Darwin from his camp in the mists 12,000 feet high, was very 'Fuegian': he was writing on his knee on top of a great rock 'with a little Tent 8 ft by 6 over me & a blazing fire in front; still the ground is sodden and I cannot keep my feet warm':

> What a real pleasure I now find in rereading your letters & killing my time with this gossip. I am above the forest region, amongst grand rocks & such a torrent as you see in Salvator Rosa's paintings vegetation all a scrub of rhodos. with Pines below me as thick & bad to get through as our Fuegian Fagi on the hill tops, & except the towering peaks of P.S. [perpetual snow] that, here shoot up on all hands there is little difference in the mt scenery – here however the blaze of Rhod. flowers & various colored jungle proclaims a differently constituted region in a naturalists eye & twenty species here, to one there, always are asking me the vexed question, where do we come from?

Hooker's mind went on turning over the great issues, as he waited within striking range of the Tibetan frontier. He wrote again to Darwin in September 1849, from 'Thibet frontier/Camp/Sikkim Himal'; he had been reading Darwin's *Geology of South America* with profit and pleasure – 'How you would have laughed could you have seen me perusing it with avidity up at 16000 ft. . . . Lying in bed huddled up in blankets with the smallest possible tip of one finger exposed to turn the pages, the skin off both nose & cheeks & holding the book with my hand cased in the blanket so awkwardly that it ever & anon bumped on to my face & deranged my spectacles.' But even in these conditions, he settled down to answer the inevitable questions which any letter from Darwin contained.

Every obstacle, bar violence, had been placed in Hooker's way.

Bridges were destroyed; food refused; false frontiers, marked by two pieces of stick with a little worsted strung between them, were thrown across his path. On one occasion he had to put his dog, Kinchin, on guard and sit with his gun beside him to support his argument. He bore no grudge against the officials who obstructed him, who were transparently acting under duress. 'Why,' they asked him, 'should you spend weeks on the coldest, hungriest, windiest, loftiest place on the earth, without even inhabitants?' 'Have you not got all the plants and stones you want? You can see the sun much better with those brasses and glasses lower down.' In the end, his persistence wore them down, and they agreed to accompany him to the true frontier. Higher up the pass he sat on a Chinese rug inside a tent, while salted and buttered tea was prepared, on his way to the Kongra Lama summit, which he estimated at 15,745 feet. For two hours there he wandered happily around, gathering forty kinds of plant, gazing across the watershed of the Himalayas at the barren sweep of land before him, the 'howling wilderness' of the high plateau of Tibet; while at his shoulder rose the great amphitheatre of rock and snow under Kinchinjhow, with an ice face of 4,000 feet, 'a great blue curtain reaching from heaven to earth'. After walking thirteen miles, he was glad to ride back by moonlight on a pony, led by a boy, while he reflected on the events of a day he had looked forward to for so long: to stand on the threshold of Tibet. Now all obstacles were surmounted, and he was returning laden with materials for extending the knowledge of a science which had formed the pursuit of his life: could it be wondered at that he felt proud?

After a month's interval, and with new supplies of rice from Darjeeling, Hooker was off again, to explore the Donkia pass, another of the routes between Sikkim and Tibet. He was in his element. One of the officials grew 'intensely disgusted' with his determination. After exhausting his persuasions, threats and warnings about snow, wind, robbers, starvation and Tibetan sepoys, he turned back, leaving Hooker 'truly happy for the first time since quitting Dorjiling': 'I had now a prospect of uninterruptedly following up my pursuits at an elevation little below that of Mont Blanc, surrounded by the loftiest mountains, and perhaps the vastest glaciers in the globe; my instruments were in perfect order, and I saw around me a curious and varied flora'. He ascended the pass twice – 18,466 feet above sea level – and in fact crossed over into Tibet, marvelling at the boundless prospect stretching out before him. From one of the cairns he extracted a slab of slate, inscribed in Tibetan 'Om Mani Padmi om', which Meepo, the Rajah's representative, allowed him to take away 'as the reward of

17. The summit of the Forked Donkia, with 'Goa' antelopes.

my exertions'. He camped lower down the valley, sometimes accompanied by passing groups of Tibetans and their yaks. The Tibetan dogs howled, and the yaks, inquisitive and bad sleepers, would push their muzzles under his tent flaps, waking him up with a snort and a blast of hot moist breath: he slept with a heavy tripod by his side, to poke at intruders. From this base, he made repeated attempts to climb Donkia and Kinchinjhow, on one occasion reaching between 19,000 and 20,000 feet, and once saw 'the spectre of the Brocken', when his own shadow was projected on to a bank of mist that rose above the precipices on whose crest he stood, while his head was surrounded with a brilliant circular glory or rainbow: the apotheosis of the naturalist. He had climbed higher than Humboldt on Chimborazo, higher than any other European before him.

But Hooker was not quite satisfied: he was a perfectionist. Campbell had arrived in the area, armed with authority to negotiate an excursion into Tibet. With Campbell and his porters, Hooker went back up the Kongra Lama, where they were met by a detachment of Tibetan troops. Not liking the way the negotiations were apparently going, Hooker took off into the mist; Campbell eventually appeared to join him, after securing agreement. Campbell's porters had not become acclimatised, and fell ill in relays, much to Hooker's delight, since that meant more time to take sightings and botanise. Eventually they moved east to the Cholamo lakes, and came back into Sikkim by the Donkia pass. By this time everyone, apart from Hooker, seemed 'quite knocked up', and he had to do a lot of encouraging and passing round of the brandy bottle. He had become used to the altitude, though on his first trip, he confided to Darwin, he had gone on retching for hours even at 15,000 feet.

They travelled south, Hooker still collecting plants and seeds, descending swiftly from temperate to tropical flora, and arrived in November at Tumloong, where the Rajah was in residence. There were some ominous signs of rudeness, but nothing to prepare for the incident which followed a trip to yet one more pass, the Chola.

> We went into the hut, and were resting ourselves on a log at one end of it, when, the evening being very cold, the people crowded in; on which Campbell went out, saying, that we had better leave the hut to them, and that he would see the tents pitched. He had scarcely left, when I heard him calling loudly to me, 'Hooker! Hooker! the savages are murdering me!' I rushed to the door, and caught sight of him striking out with his fists, and struggling violently; being tall and powerful, he had already prostrated a few, but a host of men bore him down, and appeared to be trampling on him; at the same time I was myself seized by eight men, who forced me back into the hut, and down on the log, where they held me in a sitting posture, pressing me against the wall; here I spent a few moments of agony, as I heard my friend's stifled cries grow fainter and fainter. I struggled but little, and that only at first, for at least five-and-twenty men crowded round and laid their hands upon me, rendering any effort to move useless; they were, however, neither angry nor violent, and signed to me to keep quiet.

Campbell, with whom the quarrel principally lay, had been bound hand and foot, and even mildly tortured, by having the cords twisted round his wrists with a bamboo-wrench; when Hooker saw him he was limping and had a black eye, but was otherwise looking stout and confident. Hooker expostulated that they could not play 'fast and loose with a British subject'. Campbell was a hostage, in a complicated struggle for power between the Dewan and the Rajah; the Dewan clearly hoped to use Campbell somehow to extract political advantage. By refusing to give an inch, Hooker and Campbell slowly improved the situation; Hooker was allowed to join Campbell in the small bamboo hut he had been confined in. But they were not permitted to communicate with Darjeeling, and rumours were apparently flying around that the two had been killed, and that the Rajah had sent for 50,000 Tibetan soldiers who were marching over the passes to sack Darjeeling and drive the English out of Sikkim. Hooker was eventually permitted to write to the Governor-General, who responded vigorously. A dispatch was sent to the Rajah, 'such as the latter was accustomed to receive from Nepal, Bhotan, or Lhassa, and such as alone commands attention from these half-civilized IndoChinese, who measure power by the firmness of tone adopted towards them'. At last, the smack of firm government: troops were moved north. As November dragged into December, the Dewan's confidence began to subside. Finally Hooker and Campbell were escorted to Darjeeling by an armed guard; but although from time to time the sepoys brandished their weapons menacingly, it was all only play-acting. The Rajah ended up by losing some territory; and the Dewan was disgraced and turned out of his office, forbidden from entering Tibet. The Tibetans were furious that they had been implicated in the dispute. Hooker clearly thought that a punitive expedition should have invaded and occupied the country: the naturalist had turned into an agent of empire. But for Hooker the botanist, the Himalayan journey had exceeded his 'most sanguine expectations', even though some of his collection had been destroyed. He had, too, been able to survey the whole country and make a map – 'and Campbell had further gained that knowledge of its resources which the British Government should all along have possessed, as the protector of the Rajah and his territories'.

Nothing could quite match that last expedition for excitement, though he was eager for more, and his friend Dr Thomas Thomson was waiting to join him. The Borneo venture was cancelled, and Hooker was granted £300 for a further year's botanising in India. Thomson had just been in the north-west Himalayas. Hooker was keen to spend more time in Nepal, and trek to Kathmandu, but his chief Nepalese contact, Jung Bahadoor, was going to be in England, and did not think Hooker would be safe in his absence. So Hooker and Thomson compromised on a trip through the Khasia mountains, in Assam. Botanically exciting, this was not nearly so enjoyable: the Khasians were not a patch on the Lepchas, 'sulky intractable fellows', according to Hooker. All the same, they collected more than 2,500 species – 200 men's loads. Seven loads of this was the azure orchid, *Vanda caerulea*, all destined for Kew, though very few specimens reached England alive. Hooker gives in a footnote an interesting insight into the commercial dimension. 'A gentleman who sent his gardener with us to be shown the locality, was more successful; he sent one man's load to England on commission, and though it arrived in a very poor state, it sold for £300, the individual plants fetching prices varying from £3 to £10. Had all arrived alive, they would have cleared £1000.' Suddenly, the image of the solitary plant-hunter is shattered. The place was swarming with them, he admitted in a letter to his father: 'What with Jenkins' and Simon's collectors here, twenty or thirty of Falconer's, Lobb's, my friends Raban and Cave and Inglis's friends, the roads here are becoming stripped like the Penang jungles, and I assure you for miles it sometimes looks as if a gale had strewed the road with rotten branches and Orchidae. Falconer's men sent down 1000 baskets the other day, and assuming 150 at the outside as the number of species *worth cultivating*, it stands to reason that your stoves in England will still be stocked.' Hooker steered away from the plundered orchids, and concentrated on more elusive species such as palms: the great Palm House at Kew had been completed in 1848.

Hooker's trip, exciting enough at the time, assumes in retrospect an unmistakably imperial dimension, part of a slow but inexorable process of domination and annexation. Darjeeling had been ceded to the British as recently as 1835, as a gesture of friendship on the part of the Maharajah to the Governor-General, in order, ostensibly, to create a healthy hill station 'for the purposes of enabling the servants of his Government suffering from sickness to avail themselves of its advantages'. Campbell, installed as the superintendent, among other initiatives introduced seeds and tea-plants from China – and Chinese horticulturalists to tend them – into the district in 1841: it did not take many years for Darjeeling to be transformed from a sanatorium to a tea-planting centre. Hooker, the innocent naturalist, collecting plants and taking measurements, made a very thorough exploration of the Sikkim valleys and passes, steadfastly ignoring polite and impolite requests to stop, wholly confident of his rights. In his *Himalayan Journals* he speculated that the Dewan Namgay, whose energy he recognised, might yet cut a figure in Bhutan, if not in Sikkim itself: Namgay bounced back as predicted, and in 1861 the Lieutenant Governor of Bengal invaded Sikkim with 2,600 riflemen who forced their way to Tumlong, the capital. The result was the final expulsion of Namgay, a hefty fine, and yet more territory handed over. The spectacled plant-collector in his tartan shooting-jacket, so innocently writing botanical notes in his journal, was the advance guard of the British Raj.

Hooker could also claim to have carried out pioneering field-work as a mountaineer. No European had penetrated so many of the valleys which lead to Kinchinjunga, the world's third highest mountain after Everest and K2. Although Saussure had ascended Mont Blanc, the great age of mountain climbing in the Swiss Alps had not fully begun; here was Hooker scrambling about at a level higher than any European mountain, without any specialist equipment, peering through home-made anti-glare devices made from Mrs Campbell's veil, without experienced guides, and obviously rather disappointed when he did not succeed in reaching the great glaciers or climb higher on the snow-covered peaks.

Hooker had proved a superb plant-collector, with an eye for everything from the smallest lichen to the forest trees. But his scientific mind was all-embracing, from the great land formations he had trekked amongst, the moraines and glaciers and mountains, to the people who lived in their valleys and passes, the Lepchas and Tibetans; and he thought brilliantly in generalised as well as specific terms. While he was still in India, he was correcting the first folio volume of *The Rhododendrons of Sikkim-Himalaya*, edited by his father, and working up his map of Sikkim, each in their own way testimony to his accuracy of detail. Over the next decade he continued to publish, and to liaise closely with Darwin, slowly adjusting and testing his own views about species. When Hooker wrote the introductory essay to his *Tasmanian Flora* shortly before *The Origin of Species* came out, it was an application of the principles of natural selection to a particular country's flora. When Lyell published his *Antiquity of Man* in 1863, Hooker was a little embarrassed to find himself given such a large share of the credit for '*establishing*, though not *originating*', the Darwinian theory.

6 *Wallace and the King Bird of Paradise*

WALLACE HAD SPENT A BUSY TIME in the year since his return to England in October 1852. He had sold and distributed the remnants of his collections, and worked fast on his *Travels on the Amazon and Rio Negro*, which was published in the autumn of 1853 and well received. He attended meetings of the learned societies, where he once heard Huxley speak 'with wonderful power'. He had not yet met Darwin, though the two would soon be in correspondence. The *Travels* served as Wallace's passport, and Darwin would enlist him as a valuable addition to his worldwide network. Wallace's chief object for his second expedition was 'the investigation of the Natural History of the Eastern Archipelago in a more complete manner' than had ever before been attempted; but, he added in his submission to the Royal Geographical Society, he would also 'pay much attention to Geography': he took with him the astronomical and meteorological instruments necessary to calculate the latitude, longitude and height above sea level of all his stations. His plan was to make his headquarters at Singapore, and to visit 'in succession Borneo, the Philippines, Celebes, Timor, the Moluccas and New Guinea . . . remaining one or more years in each as circumstances may determine'. Apart from the Philippines, he faithfully carried out this formidable programme. After one false start – the Royal Navy's brig *Frolic*, on which he had been allocated a free berth, was diverted to the Crimea – Wallace and a young assistant, Charles Allen, sailed to Alexandria aboard a P & O mailboat, moved up to Cairo by barge, across to Suez by horse and carriage, and then on via Aden, Galle and Penang to reach Singapore on 20 April 1854. Wallace would be away from England for almost eight momentous years.

By the time Wallace returned in 1862, the course of science had altered beyond recognition. Although the full implications of natural selection would take some years to be digested, with Wallace playing a major role in interpreting Darwin's *Origin of Species*, the scientific framework had been revolutionised; and Wallace himself had been largely instrumental in deciding the

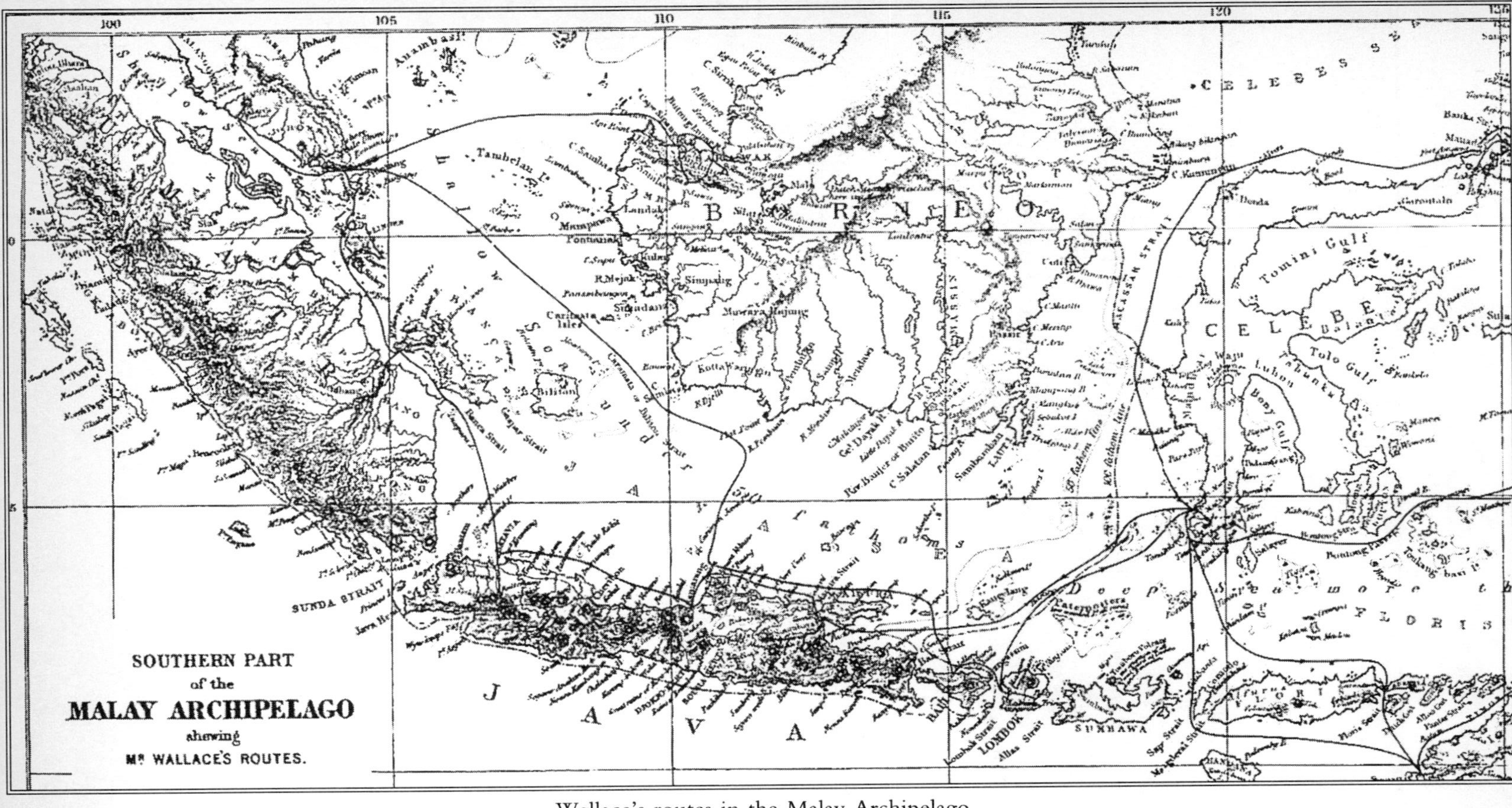

Wallace's routes in the Malay Archipelago

timing. Two significant papers by Wallace spurred Darwin into going public. These were 'On the Law which has regulated the Introduction of New Species' (Sarawak, February 1855) and, most crucially, 'On the Tendency of Varieties to Depart Indefinitely from the Original Type' (Ternate, February 1858). Meanwhile, Wallace had pursued his systematic, exhausting, exhilarating and sometimes dangerous vocation, moving from island to island and locality to locality, and at regular intervals dispatching back to England his hard-won collections, some to sell, and some to keep for his own private collection and future study. All the while he wrote and annotated and recorded. The sheer scale of his industry is astonishing. The specimens he collected had to be identified, labelled, preserved and packed. There were animals to skin, and reptiles to stuff. There were his journals to keep. There were long letters to write, some to do with detailed arrangements about his collections, some to ensure money and key supplies reached him in anything up to six months' time: he never had the benefit of the free postal arrangements enjoyed by Hooker in India, or the semi-official back-up of the curator of a colonial botanic garden; and, in the midst of all this, he still had the intellectual energy to conduct a lively correspondence with other naturalists and scientists, such as Darwin or his old friend Bates, and to grapple with the fundamental questions about the origins and evolution of man.

Reading *The Malay Archipelago*, which he delayed writing for six years after his return to England, but which is based on the journals written day by day, one is struck by Wallace's acute interest in the peoples among whom he travelled. This trait is true, also, of *Travels on the Amazon*, particularly in the passages on the Indians of the Uaupés river. But there Wallace seems simply to express his characteristic warmth and humanity. In *The Malay Archipelago*, his careful recording and his analyses seem more systematic, as though his observations are part of a general scientific and philosophical enquiry. One major result of his expedition was clarification of the zoological geography of the Malay Archipelago, a discovery commemorated by the description 'Wallace's Line', dividing the Asiatic from the Australian fauna. It is clear that he is always thinking not just in terms of animal species, but of human races:

> As bearing upon this question it is important to point out the harmony which exists between the line of separation of the human races of the Archipelago and that of the animal productions of the same country, which I have already so fully explained and illustrated. The dividing lines do not, it is true, exactly agree; but I think it is a remarkable fact, and something

> more than a mere coincidence, that they should traverse the same district and approach each other so closely as they do.

'The Races of Man' forms Wallace's last chapter, and perhaps his most speculative. It has a boldness about the place of man in the natural world which Darwin's laconic aside in *The Origin of Species* famously avoided.

When *The Malay Archipelago* was eventually published in March 1869, it was lavishly illustrated; and the choice of frontispiece is highly significant. The book's subtitle is: 'The Land of the Orang-Utang and the Bird of Paradise', and the frontispiece, by Joseph Wolf, pleased Wallace enormously. The title is 'Orang-Utang Attacked by Dyaks', a neat reversal of what might, at first glance, be taken for an aggressive Orang pouncing on a passing Dyak. The orangs, avid for durian fruit, were up against stiff competition in the forests of Borneo, and were pursued relentlessly:

> A few miles down the river there is a Dyak house, and the inhabitants saw a large orang feeding on the young shoots of a palm by the river-side. On being alarmed, he retreated toward the jungle which was close by, and a number of the men, armed with spears and choppers, ran out to intercept him. The man who was in front tried to run his spear through the animal's body, but the mias [orang] seized it in his hands, and in an instant got hold of the man's arm, which he seized in his mouth, making his teeth meet in the flesh above the elbow, which he tore and lacerated in a dreadful manner. Had not the others been close behind, the man would have been more seriously injured, if not killed, as he was quite powerless, but they soon destroyed the creature with their spears and choppers.

Wallace duly took possession of the skeleton of this particular animal: it yielded him a very fine skull.

Wallace's interest in the orang, or the mias, was complex. He wanted, he claimed, to observe its behaviour, in its natural habitat. 'One of my chief objects in coming to stay at Simunjon was to see the orang-utan (or great man-like ape of Borneo) in his native haunts, to study his habits, and obtain good specimens of the different varieties and species of both sexes, and of the adults and young animals.' There were so few orang skins, skeletons and skulls in Europe that he knew there would be a ready market for them; the successful capture of a good set of mature orangs would help to finance his expedition.

The orang phase came relatively early in his journey, during a visit to Borneo which began with the first of two Christmas stays with Sir James Brooke, the first white Rajah of Sarawak. Brooke

18. Orang-utan attacked by Dyaks. (*The Malay Archipelago*)

was himself a naturalist, and had written a paper on the orang in 1842. According to Brooke's first biographer, his private secretary Spencer St John, Wallace was even then 'elaborating in his mind the theory which simultaneously was worked out by Darwin – the theory of the origin of species; and if he could not convince us that our ugly neighbours, the orang-outangs, were our ancestors, he pleased, delighted, and instructed us by his clever and inexhaustible flow of talk – really good talk. The Rajah was pleased to have so clever a man with him as it excited his mind, and brought out his brilliant ideas.' At Kuching, or in Brooke's mountain cottage retreat, the discussions were always either philosophical or religious.

In March 1865, with the rainy season over, Wallace moved away from familiar and well-described territory to the newly opened coal mines on the Simunjon river. It is something of a shock to read that, after one brief observation of an orang, the ruthlessness of Wallace's collecting instinct took over. The next encounters all lead inexorably to an orang's death, often after several shots, with lengthy pursuits and difficult retrievals. Wallace once offered four Chinese a day's wages each to cut down a large tree in which a wounded orang had made a nest: they refused, and Wallace, not wanting to pay over the odds so soon when he planned to be in the area for several months, declined to bargain; three months later, two Malays climbed the tree for a dollar to secure the dried remains for him.

While pursuing his quarry, Wallace was also noting a great many details about the orang, even if his approach was in stark contrast to that of a twentieth-century naturalist. He also acquired an infant orang. After shooting a full-grown female, he found a very young orang lying face down in the bog. He cleaned it up and took it back to camp, feeding it with rice-water from a bottle with a quill in the cork. Wallace's description deliberately emphasises the similarities in behaviour to a human. 'When I put my finger in its mouth, it sucked with great vigor, drawing in its cheeks with all its might in the vain effort to extract some milk, and only after persevering a long time would it give up in disgust, and set up a scream very like that of a baby in similar circumstances. . . . I fitted up a little box for a cradle, with a soft mat for it to lie on, which was changed and washed every day, and I soon found it necessary to wash the little miss as well. After I had done so a few times, it came to like the operation, and as soon as it was dirty would begin crying'. Wallace gave it a comfort cloth, made a short ladder to help it exercise, and began to feed it on solids. He compared its development to that of a young hare-lip monkey, which

19. Female orang-utan.

he soon acquired as a suitable companion for it, and noted how it gradually began to learn to run about on its own, just as a young human child would do. Sadly, it did not grow as it should, and after three months became seriously ill – the symptoms 'exactly those of intermittent fever' – and died. 'I much regretted the loss of my little pet, which I had at one time looked forward to bringing up to years of maturity, and taking home to England.' Moving from this attempt to raise a young orang, as he might have nurtured an orphan, Wallace then catalogues his collection, and concludes the section by summarising his observation of the orang's habits, and making the following speculation:

> It is very remarkable that an animal so large, so peculiar, and of such a high type of form as the orang-utan, should be confined to so limited a district – to two islands, and those among the last inhabited by the higher Mammalia. . . . When we consider, further, that almost all other animals have in earlier ages been represented by allied yet distinct forms . . . we have every reason to believe that the orang-utan, the chimpanzee, and the gorilla

> have also had their forerunners. With what interest must every naturalist look forward to the time when the caves and tertiary deposits of the tropics may be thoroughly examined, and the past history and earliest appearance of the great man-like apes be at length made known.

In a paper published in 1856, 'On the Habits of the Orang-Utan of Borneo', aimed at a more specialised readership, Wallace made the connection with man even more obvious:

> One cannot help speculating on a former condition of this part of the world which should give a wider range to these creatures, which at once resemble and mock the 'human form divine,' – which so closely approach us in structure, and yet differ so widely from us in many points of their external form.

In the path of Wallace's speculation and prediction would lie the discovery, at the end of the century, of 'Java Man'. Meanwhile, in advance of his own Ternate paper, he points as clearly as perhaps he dared to the common ancestry of man and ape. A few years later Huxley made the connections even more explicit in his 1863 collection of essays, *Man's Place in Nature*, which incorporates a considerable number of Wallace's observations on the orangs.

More immediately and pragmatically, the orangs had to be sold. The *Water Lily* sailed from Singapore on 5 March 1856 with a cargo which included two casks containing five orang skins in arrack, and a box which held sixteen orang skulls and two skeletons. The skins were insured for £50, though Wallace hoped for £250 or even £300 for the series – he instructed his agent to consult Professor Owen and George Waterhouse, the curator of the Zoological Society. There was also, disconcertingly, a human skull – 'for J.B. Davis Esq., Shelton, Staffordshire, who will send for it'. Much to Wallace's later annoyance, H.M. Customs tried to make Stevens pay duty on the arrack: 'My fishes from Para, Bates fishes etc. never paid any duty and I think that precedent should be urged as had I known I should have used brine instead of spirit. Surely my character and yours should be sufficient . . . '. In addition to the orangs, there were 7,000 insects – 5,000 for sale and 2,000 'private' – 60 bird skins, mammal skins, reptiles and shells. Many of these items needed identification: 'The Mammals as before, sell, keeping back for me a series of the squirrels & of all others which Mr Waterhouse cannot name.' Meanwhile, Wallace was effectively confined to Singapore for another month: 'As it will be three months before I get money from you and after my expenses here and the necessary outfit of clothes, ammunition &

other necessaries which I *must* now get, I shall have little enough to pay my passages to Celebes & live for two months.'

He did not leave Singapore for Bali until 24 May – six months 'utterly lost and at great expense'; people never reckoned such factors, he complained, when estimating the profits of collectors. He spent some time with Thomas Lobb, the great orchid-hunter who worked for the commercial nurseryman Veitch. Lobb did not think much of the Moluccas for plants, but Wallace had great faith in them for insects and birds; and if he could reach the bird of paradise country, the Aru Islands, he should be able to prepare good specimens of those gorgeous birds, 'one of the greatest treats I can look forward to'. Meanwhile, he looked forward to receiving his mended spectacles, and a second pair as a reserve. There was still no news about the orangs. He wrote again from Lombok in August: collecting was still slow work – 'here there are nothing but dusty roads, and paddy fields for miles around producing no insects or birds worth collecting'. But he could still send off a case with 300 birds and 465 insects, including a domestic duck for Darwin, who might also like to take the jungle cock, 'which is often domesticated here and is doubtless one of the originals for the domestic breed of poultry'. By September 1856 he was in the Celebes, at Macassar – the land was 'naked' and 'uninviting', very unlike the Amazon, where there were always good forest collecting grounds within minutes of any town. He was planning to take a small house near the forest for a base, and beginning to realise that he needed to go to much more remote locations.

The most productive and happy period Wallace spent during his eight-year expedition was on the Aru Islands, which he made his base from January to July 1857. These islands, off the south coast of New Guinea, and about a thousand miles east of Macassar, provided him with his most extensive and valuable collection: 9,000 specimens, of about 1,600 distinct species, whose sale financed his researches for a further five years. These months, too, led up to his great breakthrough in theory, the Ternate paper of February 1858, and helped to confirm his growing convictions about the biogeography of the area. Wallace's vivid engagement with the islands emerges from his writing, which characteristically placed the human inhabitants first in his enthusiastic summary:

> I had made the acquaintance of a strange and little-known race of men; I had become familiar with the traders of the far East; I had revelled in the delights of exploring a new fauna and flora, one of the most remarkable and most beautiful and least-known in the world; and I had succeeded in the main object for which I had undertaken the journey – namely, to obtain fine specimens

> of the magnificent birds of paradise, and to be enabled to observe them in their native forests.

The bird of paradise was both a precious rarity and a symbol, for Wallace. He had only been on the main island for a few days, with his collecting programme hampered by wet weather, when his boy Baderoon returned with a specimen which repaid him for months of delay and expectation:

> It was a small bird, a little less than a thrush. The greater part of its plumage was of an intense cinnabar red, with a gloss as of spun glass. On the head the feathers became short and velvety, and shaded into rich orange. Beneath, from the breast downward, was pure white, with the softness and gloss of silk, and across the breast a band of deep metallic green separated this colour from the red of the throat. Above each eye was a round spot of the same metallic green; the bill was yellow, and the feet and legs were of a fine cobalt blue, strikingly contrasting with all the other parts of the body. Merely in arrangement of colours and texture of plumage this little bird was a gem of the first water, yet these comprised only half its strange beauty . . .

He goes on to describe two more ornaments, the breast fans and the spiral-tipped tail-wires, 'altogether unique, not occurring on any other species of the eight thousand different birds that are known to exist upon the earth', which made this species 'one of the most perfectly lovely of the many lovely productions of nature'. Wallace's transports of delight amused his Aru hosts, who saw nothing more in the 'burong rajah' than a European would in a robin or goldfinch.

To have obtained a good specimen of the King Bird of Paradise, hitherto described by Linnaeus only from mutilated skins, fulfilled one of the objects of Wallace's long journey. Very few Europeans had ever seen 'the perfect little organism'. 'The emotions excited in the minds of a naturalist who has long desired to see the actual thing which he has hitherto known only by description, drawing, or badly-preserved outer covering, especially when that thing is of surpassing rarity and beauty, require the poetic faculty fully to appreciate them.' Wallace, inherently modest, clearly doubted his ability to do justice to this 'thing of beauty', but his very consciousness of inadequacy adds a dimension to his description.

As so often, too, the discovery of some special bird or animal prompts him to place its existence within a wider context, in a passage which emphasises the connectedness of all organisms, including 'civilised' man:

20. The 'Twelve-wired' and the 'King' Birds of Paradise.

> It seems sad that on the one hand such exquisite creatures should live out their lives and exhibit their charms only in these wild inhospitable regions, doomed for ages yet to come to hopeless barbarism; while, on the other hand, should civilized man ever reach these distant lands, and bring moral, intellectual, and physical light into the recesses of these virgin forests, we may be sure that he will so disturb the nicely-balanced relations of organic and inorganic nature as to cause the disappearance, and finally the extinction, of these very beings whose wonderful structure and beauty he alone is fitted to appreciate and enjoy. This consideration must surely tell us that all living things were *not* made for man. Many of them have no relation to him. The cycle of their existence has gone on independently of his, and is disturbed or broken by every advance in man's intellectual development; and their happiness and enjoyments, their loves and hates, their struggles for existence, their vigorous life and early death, would seem to be immediately related to their own well-being and perpetuation alone, limited only by the equal well-being and perpetuation of the numberless other organisms with which each is more or less intimately connected.

If Wallace's line of thinking occasionally ascribes human passions, loves and hates to the birds of paradise, the more radical implication is that man's cycle of existence is directly comparable to that of other organisms, and interdependent with them. Wallace also, with what now seems devastatingly accurate foresight, predicts the inevitable destruction which will accompany the nicely balanced relations of organic and inorganic nature within the virgin forests of the island. His vision of paradise is accompanied by the melancholy certainty that it is at best threatened. Meanwhile, he was able to go into the forest and observe the habits of other species of paradise bird. From his hut, before dawn, he woke to their cries as they went to seek their breakfast; and, a few weeks later, was lucky enough to see as many as twenty male birds at once in their 'dancing-parties, assembled in trees which had an immense head of spreading branches and large but scattered leaves, which gave a clear space for the birds to play and show off their plumes'.

As he had done in the Amazon, Wallace was moved to extend his admiration for the natural richness surrounding him to the Aru islanders themselves. In fact both his journals, and the book which was based on them, give the impression of a man who is as interested in the people as in the birds and insects he was studying; interested in them as individuals, and as species. Which people were Malay, and which Papuans? Where did the dividing line fall?

21. Natives of Aru shooting the Great Bird of Paradise.

Did it fall in the same place as the line dividing the fauna? And behind these questions lies the key problem: what features or qualities or factors lead to the healthy survival and development of a species, or a variety? It is hard to escape the conclusion that Wallace was investigating the species man, man in Aru, man in Gilolo, at the same time that his mind was engaged on the broader question which culminated in his paper, 'On the Tendency of Varieties to Depart Indefinitely from the Original Type'. There were, in all, some eight papers and closely argued letters which stemmed directly from these crucial months in Aru.

It is worth putting the journal entry about the Aru islanders alongside the slightly smoother, more restrained version which appeared almost twelve years later. First, the published text: 'Here, as amongst most savage people I have dwelt among, I was delighted with the beauty of the human form – a beauty of which stay-at-home civilized people can scarcely have any conception. What are the finest Grecian statues to the living, moving, breathing men I saw daily around me? The unrestrained grace of the naked savage as he goes about his daily occupations, or lounges at his ease, must be seen to be understood; and a youth bending his bow is the perfection of manly beauty.' The journal, entry 83, reads: 'Here, as among the Dyaks of Borneo and the Indians of the Upper Amazon I am delighted with the beauty of the human form, a beauty of which stay at home civilized people can never have any conception. What are the finest grecian statues to the living moving breathing forms which everywhere surround me. The unrestrained grace of the naked savage as he moves about his daily occupations or lounges at his ease must be seen to be understood. A young savage handling his bow is the perfection of physical beauty. Few persons feel more acutely than myself any offence against modesty among civilized folk, but here no such ideas have a moment's place; the free development of every limb seems wholly admirable, and made to be admired.' Wallace has controlled his admiration in the revised version, reining back the civilised/savage contrast, and, perhaps to avoid a charge of offending modesty, centring his praise on the 'manly': he changed 'forms' to 'men', cut the last sentence, and moved swiftly on to a disclaimer: 'The women, however, except in extreme youth, are by no means so pleasant to look at as the men.'

Wallace was never blinkered in his appreciation of a 'natural' way of life. He recognised that it imposed a considerable burden, mostly carried by the women, who suffered from the combination of hard work and very early marriages. Wallace was, too, extremely aware of variations in the habits and diets of the different people he lived among; he was a strong believer in a work ethic, and critical of the local diet in this instance – the Aru people had no regular supply of rice, *mandiocca*, maize or sago; they were too dependent on fruit and vegetables, and so subject to skin diseases and ulcers; the men in Wokam, the village he first stayed in, did not hunt regularly. Man, he concluded, is not able 'to make a beast of himself with impunity, feeding like the cattle on the herbs and fruits of the earth, and taking no thought of the morrow'. He needed some farinaceous product capable of being stored and accumulated, to give him a regular supply of wholesome food. In Wanumbai, his second base, there was a much more animated and energetic way of life. The house he lived in there contained four or five families:

> They keep up a continual row from morning to night, – talking laughing shouting without intermission; not very pleasant, but I take it as a study of national character and submit. My boy Ali says "Banyak quot bitchara Orang Arru" (The Arru people are very strong talkers). . . . All the men and boys are expert archers never stirring out without their bows and arrows. They shoot all sorts of birds as well as pigs and kangaroos occasionally, which gives them a pretty regular supply of meat with their vegetables. The result of this better living is superior healthiness, well made bodies and generally clear skins.

In designating Aru as the home of Papuan rather than Malay, or transitional, peoples, Wallace was challenging accepted contemporary theory. He seems to be implying that races are not permanent varieties. The lifestyle of one village differs from its equivalent on the north of the island: the diet, the habits, even what Wallace would later describe as the moral qualities, differ, making one community healthier, more likely to prosper, than the other in certain sets of circumstances. Like the size of the beaks of Darwin's finches, the physical skill of a young Aru hunter with a bow may make all the difference between survival and development, and extinction.

Inspired by his four-month stay, Wallace left before the monsoon struck, and re-established himself at Macassar, where he spent a month fully occupied in sorting, cleaning and packing up his Aru collection. He had his guns repaired and, with a fresh supply of pins and arsenic, gathered his resources to investigate a new collecting ground in the Celebes. From there, he took a Dutch mail-ship for Amboyna and Ternate, in the Moluccas. He had not, however, finished with his birds of paradise. In October 1858 he hired a small boat and explored Bacan: 'Luckily it was fine

22. Wallace's house on Waigiou, off the north-west tip of Papua New Guinea (now Irian Jaya).

weather,' he wrote to Stevens, 'or a hundred miles at sea with no means of cooking and only room for one day's water would have been more than unpleasant.' The Dutch were working the coal deposits, and a good road to the mines gave him easy access to the forest. He was excited about the locality – in butterflies he took an admittedly imperfect specimen of a glorious new species 'very like Ulysses but distinct and even handsomer'; he rejoiced in its beauty, and equally enthusiastically worked out its value in money: 'Perfect specimens of this must be five pounders.' But he already had the finest and most wonderful bird in the island: 'I had a good mind to keep it a secret but I cannot resist telling you. I have got here a new Bird of Paradise! of a new genus!! quite unlike anything yet known, very curious and very handsome!!! When I can get a couple of pairs I will send them overland to see what a new Bird of Paradise will really fetch. I expect £25 each!' He thought it the greatest discovery he had yet made; and when it reached the British Museum it was duly named by Gray 'Wallace's Standard-wing'.

23. 'Wallace's Standard Wing', a new Bird of Paradise.

Between these two glimpses of paradise, on Aru and Bacan, Wallace had conceived the theory of natural selection, and written the paper which influenced the course of science and effectively transformed his own career and standing, from a collector to a scientist. The same letter which announced his new bird of paradise contained, after a mild complaint about a set of pins which were no good for the longer-legged beetles, the following low-key request:

> An Essay on Varieties which I sent to Mr Darwin has been read to the Linnaean Society by Dr Hooker and Mr C. Lyell on account of an extraordinary coincidence with some views of Mr Darwin, long written but not yet published and which were also read at the same meeting. If these are published I dare say Mr Kippish will let you have a dozen copies for me. If so send me 3 and of the remainder send one to Bates, Spencer and any other of my friends who may be interested in the matter and who do not attend the Linnaean.

A sense of mystery, or, as E.M. Forster might have described it, muddle, swirls around the scientific events of 1858. The story has often been told, but, in the absence of some key letters and documents, there is an inviting gap for speculation, even for a conspiracy theory that has Darwin frantically reshaping his paper after receiving Wallace's bombshell from Ternate. The truth is almost certainly more mundane. But there is a poetic charge to the events, and to the sharp contrast in style between the two men, between their backgrounds, methods and circumstances. In a sense, this episode forms the ultimate triumph of the nineteenth-century scientific traveller: the discovery of the new, overarching pattern or outline of the great tree, or bush, of life, to replace the laborious accumulation of isolated pieces turned this way and that in an attempt to discover how they fitted together.

Wallace, cut off from scientific conversation, had to conduct his discussions at long range; in addition to his regular letters to and from his agent, he wrote at length and regularly to Bates, still in the Amazon, and to Spruce, in the Andes; and, beginning in 1856, to Darwin. Darwin, writing on 1 May 1857, from Down, thanked Wallace for valuable and real encouragement in his 'laborious undertaking' – the full-scale version of his great 'Species' book. He also commended Wallace's 1855 paper, 'On the Law which has regulated the Introduction of New Species': 'I can plainly see that we have thought much alike and to a certain extent have come to similar conclusions.' Darwin agreed 'to the truth of almost every word' of the paper. He then makes crystal clear to Wallace the state of his own work: 'This summer will make the twentieth year (!) since I opened my first note-book on the question how and in what way do species and varieties differ from each other. I am now preparing my work for publication, but I find the subject so very large, that though I have written many chapters, I do not suppose I shall go to press for two years.' This letter could be read as a clear marking out of Darwin's territory, especially as he goes on to ask how long Wallace plans to stay in the Malay Archipelago: 'I wish I might profit by the publication of your Travels there before my work appears, for no doubt you will reap a large harvest of facts.' Wallace's role was firmly hinted at: an accumulator of facts, a worthy provider of domestic poultry skins, an answerer of questions: 'Can you tell me positively that black jaguars or leopards are believed generally or always to pair with black?' 'Is the case of parrots fed on fat fish turning colour mentioned in your Travels?' Yet Darwin cannot avoid edging towards the theoretical discussion Wallace clearly wished to engage in – but, characteristically, he held back from it: 'It is really *impossible* to explain my views in the compass of a letter as to causes and means of variation in a state of nature; but I have slowly developed a distinct and tangible idea – whether true or false others must judge.' The dialogue had begun; and Darwin also asked for Wallace's views on the distribution of organic beings on oceanic islands, a subject Wallace had just been investigating on the Aru Islands, and which formed the basis of his next scientific paper. Wallace's first letter had been answered in seven months. The next was answered in three, and it must have contained a leading question. 'You ask whether I shall discuss Man: I think I shall avoid the whole subject, as so surrounded with prejudices, though I fully admit that it is the highest and most interesting problem for a naturalist.' This is the clearest indication that Wallace himself must have been thinking of man as a species, as an integral part of his theorising in his extremely productive post-Aru period. Wallace's 'Law on the introduction of New Species', and his highlighting of the central issue of man, were the preliminary signals to Darwin that the point of crisis was approaching. Darwin knew, better than anyone, what the implications of natural selection were; for this, he had written his abstract, and placed it in a sealed envelope with instructions to his wife on what to do with it in the event of his death. This was what all the disagreements about transmutation were concerned with, what Huxley's fierce tussles with Richard Owen over gorilla skulls were really about: was man an animal, and subject to the laws of nature like every other organism, or a separate kind of creation, governed by some divine or 'superior' dispensation?

Wallace, shifting his base from island to island in the Moluccas, was immensely busy. But, apart from the necessary routines of his simple daily life, and his travel arrangements, he had no other distractions to cut into his thinking time. On 4 December 1857 he arrived at Ambon, where he spent a month collecting, before going on to Ternate, which he reached on 8 January. From this moment, his movements become less clear. His journals show he made a trip to the neighbouring island of Gilolo in February, collecting for a period, but also losing two weeks through illness. Neither the journals nor *The Malay Archipelago* shed much light on his precise itinerary. Years later, he described the circumstances of his Ternate paper. He was suffering from a sharp attack of fever, which forced him to lie down each afternoon during the two or three hours that the alternating cold and hot fits lasted:

> It was during one of these fits, while I was thinking over the possible mode of origin of new species, that somehow my thought turned to the 'positive checks' to increase among savages and others described in much detail in the celebrated Essay on Population, by Malthus, a work I had read a dozen years before. These checks – disease, famine, accidents, wars, &c. – are what keep down the population, and it suddenly occurred to me that in the case of wild animals these checks would act with much more severity, and as the lower animals all tended to increase more rapidly than man, while their population remained on the average constant, there suddenly flashed upon me the idea of the survival of the fittest – that those individuals which every year are removed by these causes, – termed collectively the 'struggle for existence' – must on the average and in the long run be inferior in some one or more ways to those which managed to survive. . . . Then it suddenly flashed upon me that this self-acting process would necessarily *improve the race*, because in every generation the inferior would inevitably be killed off and the superior would remain – that is, the *fittest would survive*. Then at once I seemed to see the whole effect of this, that when changes of land and sea, or of climate, or of food supply, or of enemies occurred – and we know that such changes have always been taking place – and considering the amount of individual variation that my experience as a collector has shown me to exist, then it followed that all the changes necessary for the adaptation of the species to the changing conditions would be brought about; and as great changes in the environment are always slow, there would be ample time for the change to be effected by the survival for the best fitted in every

> generation. In this way every part of an animal's organization could be modified exactly as required, and in the very process of this modification the unmodified would die out, and thus the *definite* characters and the clear *isolation* of each new species would be explained. The more I thought it over the more I became convinced that I had at length found the long-sought-for law of nature that solved the problem of the origin of species. . . . I waited anxiously for the termination of my fit that I might at once make notes for a paper on the subject. The same evening I did this pretty fully, and on the two succeeding evenings wrote it carefully to send it to Darwin by the next post, which would leave in a day or two.

This was the bombshell which landed on Darwin's doormat on 18 June 1858: a lucid, elegant statement of the theory on which Darwin had been working for twenty years, in a form clearly ready for and intended for publication, with a request that Darwin should send it on to Sir Charles Lyell. For some time Lyell and Hooker had been urging Darwin to publish his theory. Now Darwin saw their anxieties come true. His priority was 'smashed'. What, in honour, could he do about it? All his energy had gone into his huge, projected work – and he knew that those laborious researches and experiments and arguments, the amazing accumulation of facts and examples, were essential as evidence to support the theory. Yet here was an agonising moral dilemma, one he would be hard put to deal with at the best of times; five days after Wallace's letter and its enclosure arrived, his son Charles fell sick with scarlet fever, and died within days. Prostrate with grief, Darwin threw himself on the advice and help of his friends.

Lyell and Hooker were fully in touch with Darwin's thinking and research. The arrangement they made sought to be fair both to Darwin, the established and published scientific thinker, and to Wallace, the collector who had begun to write theoretical papers to add to his more focused observations on particular species and localities. They arranged to have Wallace's paper read at a special meeting of the Linnean Society on 1 July. It took its place as one of three items: extracts from Darwin's 1844 essay, part of a letter written by Darwin to Asa Gray in 1857, and Wallace's Ternate paper. Darwin stayed at home at Down, unwell, and still devastated by the death of his child. Although intense interest was aroused, according to Hooker, no discussion took place: 'the subject was too novel, too ominous, for the old school to enter the lists before armouring. . . .'

Wallace was in New Guinea, totally unaware of what might be

happening. He never complained. On the contrary, he wrote to his mother on 6 October, 'I have received letters from Mr Darwin and Dr Hooker, two of the most eminent naturalists in England, which has highly gratified me. I sent Mr Darwin an essay on a subject on which he is now writing a great work. He showed it to Dr Hooker and Sir C. Lyell, who thought so highly of it that they immediately read it before the Linnean Society. This assures me the acquaintance and assistance of these eminent men on my return home.' The speed, of course, had less to do with Hooker's and Lyell's high opinion, than the need to get something of Darwin's on the official record. Although Darwin, with the active support of Lyell and Hooker, had indeed ensured that Wallace's paper was published, the order which was chosen relegated Wallace in time, even though his was the most finished, extensive and complete statement of the three; and somehow, too, the sheer weight of titles and honours seems to put Wallace in his place: 'By Charles Darwin, Esq. F.R.S., F.L.S., & F.G.S., and Alfred Wallace, Esq. Communicated by Sir Charles Lyell, F.R.S., F.L.S. and J.D. Hooker, M.D., V.P.R.S., F.L.S. &c.' The explanation attached to Wallace's paper makes it seem an essentially private document: 'This was written at Ternate in February, 1858, for the perusal of his friend and correspondent Mr Darwin, and sent with the express wish that it should be forwarded to Sir Charles Lyell, if Mr Darwin thought it sufficiently novel and interesting'. Wallace, it is implied, was extremely lucky to have been so generously treated, though Lyell and Hooker's introductory letter is magnanimous enough to state that both Darwin and Wallace 'may fairly claim the merit of being original thinkers in this important line of inquiry'.

Darwin was relieved by Wallace's long-delayed reaction, and the generous, friendly tone of his letters to Darwin himself and to Hooker. 'Permit me to say how heartily I admire the spirit in which they are written. Though I had absolutely nothing whatever to do with leading Lyell and Hooker to what they thought a fair course of action, yet I naturally could not but feel anxious to hear what your impression would be.' Wallace was eager to know Lyell's reaction – 'somewhat staggered', Darwin informed him; Darwin thought Lyell would end by being 'perverted', while Hooker had become 'almost as heterodox as you or I'; and Hooker was by far the most capable judge in Europe. Everyone, he commented, thought Wallace's paper 'very well written and interesting'. Darwin returned as though drawn by a magnet to the question of priority, and the Linnean proceedings: 'It puts my extracts (written in 1839, now just twenty years ago!), which I must say in apology were never for an instant intended for publication, in the shade.' As Darwin concedes, Wallace's paper, by contrast, was very obviously intended for publication.

Wallace spent another four years in the Malay Archipelago, after that burst of inspiration in Ternate which precipitated natural selection into the public domain. Letters, notes, articles, continued to flow. His collecting, too, went on, as it had to. In due course, Darwin's *Origin of Species* arrived. Wallace was ecstatic. 'I know not how or to whom to express fully my admiration of Darwin's book,' he wrote to Bates. 'To him it would seem flattery, to others self-praise; but I do honestly believe that with however much patience I had worked up and experimented on the subject, I could never have *approached* the completeness of his book. . . . I really feel thankful that it has not been left to me to give the theory to the public. Mr Darwin has created a new science and a new philosophy'. Wallace was once again in Ternate. He spent a few months in Timor, then slowly made his way west, leaving the wild and savage Moluccas and New Guinea for Java, the 'Garden of the East', and Sumatra. Even the tropical forest was beginning to pall: 'a field of buttercups, a hill of gorse or of heather, a bank of foxgloves and a hedge of wild roses and purple vetches surpass in *beauty* anything I have ever seen in the tropics.' He started to enquire about arrangements for a cottage with sufficient space for his mother, as well as a study for himself and a good-sized room for his collections, and his thoughts had also turned towards marriage: 'I believe a good wife to be the greatest blessing a man can enjoy, and the only road to happiness'.

He left Singapore in February 1862, with two live specimens of the lesser bird of paradise. Having seen his monkeys and parrots perish at sea on his way back from the Amazon, he was determined to preserve these living treasures. He bought them bananas in Bombay, and bullied the Egyptian officials into allowing them to travel from Suez to Alexandria by rail. Another problem came from the P & O ships – they were so clean, that he could not find enough cockroaches to keep the birds fed. So he landed at Malta, and tracked down an excellent supply in a local bakery. With these stored in biscuit tins, he managed to sustain them on the last leg of the trip to Marseilles, and by rail through France. He arrived in London on 1 April 1862, eight years after his departure, and duly delivered the birds of paradise to the Zoological Gardens.

Wallace looked back on his travels in the archipelago as the 'central and controlling incident' of his life. He was a long time in turning his journals into a book, in spite of the repeated urgings and encouragement of Darwin. First, there were his collections to reorganise; then came the writing of papers, and Wallace's own participation in meetings of the London learned societies, as he took his place there as a major scientific thinker. There was also an unhappy love affair, which involved a proposal and a rejection; and an unfruitful application to become assistant secretary to the Royal Geographical Society: the successful candidate was his Amazon colleague Bates. In the same year, though, 1864, he began to court Mary Mitten, the eighteen-year-old daughter of a Sussex botanist; he married her the following year. It was a happy marriage; with her support, he was able to look forward to the quiet enjoyment of his collections, and to tackling the reorganisation of his journals into a book. When it was published in 1869, he wrote a warm and personal dedication to Charles Darwin. 'Not only as a token of personal esteem and friendship but also to express my deep admiration for his genius and his works.' The book was well reviewed, reprinted repeatedly, and translated into German and Dutch. Darwin told him that the dedication was something for his children's children to be proud of; the book was excellent, and at the same time pleasant to read. 'Of all the impressions which I have received from your book,' he wrote with some justice, 'the strongest is that your perseverance in the cause of science was heroic.'

Wallace, like Darwin, took care initially not to allow his more radical views on man to dominate his public and popular writing. It is possible to read *The Malay Archipelago* simply as the narrative of a natural historian; and while Wallace's rearrangement of his travels into five groups of islands – Indo-Malay, the Timor Group, the Celebes Group, the Moluccas, the Papuan Group – promotes an understanding of the scientific description and analysis, it also makes it harder to follow the evolution of his thought. Yet the subtitle, 'A Narrative of Travel, with Studies of Man and Nature', firmly places man at the centre of the enquiry, and the whole book concludes with a powerful summary, 'The Races of Man in the Malay Archipelago'. Back once more in Victorian England, in the heart of a commercial, expansionist Europe, he found a clearer perspective. Having attempted to define the various races, Wallace moves towards a sombre set of predictions, which have clear implications for a European North still bent on imperial consolidation and conquest:

> If the past history of these varied races is obscure and uncertain, the future is no less so. The true Polynesians, inhabiting the farthest isles of the Pacific, are no doubt doomed to an early extinction. But the more numerous Malay race seems well adapted to survive as the cultivator of the soil, even when its

> country and government have passed into the hands of Europeans. If the tide of colonization should be turned to New Guinea, there can be little doubt of the early extinction of the Papuan race. A warlike and energetic people, who will not submit to national slavery or to domestic servitude, must disappear before the white man as surely as do the wolf and the tiger.

There could hardly be a sharper statement of the laws of the survival of the fittest; and these are the very people whose energy and beauty Wallace admired. The qualities which promoted their well-being in the forest would ensure their destruction when confronted with the forces of European civilisation.

Wallace believed, not in orthodox Christianity, but in progress and perfectibility, in an ideal social state towards which mankind was 'tending'. His early exposure to Robert Owen's ideas may lie behind a statement such as this: 'In such a state every man would have a sufficiently well balanced intellectual organisation to understand the moral law in all its details, and would require no other motive but the free impulses of his own nature to obey that law.' This social state Wallace found, he argued, among people 'in a very low stage of civilisation'. He had lived among communities of 'savages' in South America and in the East, who had no laws or law courts except 'the public opinion of the village freely expressed'. In such communities all are equal, or more nearly equal. There were none of the wide distinctions of education and ignorance, wealth and poverty, master and servant, which are the product of European, Northern civilisation: 'there is none of that wide-spread division of labour, which, while it increases wealth, produces also conflicting interests; there is not that severe competition and struggle for existence, or for wealth, which the dense population of civilised countries inevitably creates.' Where Europe had progressed in intellectual achievements, Wallace went on to argue, it had not done so in morals: the mass of our populations 'have not at all advanced beyond the savage code of morals, and have in many cases sunk below it'.

Intellectual and material progress had occurred too fast for the European nations to absorb.

> Our mastery over the forces of nature has led to a rapid growth of population, and a vast accumulation of wealth; but these have brought with them such an amount of poverty and crime, and have fostered the growth of so much sordid feeling and so many fierce passions, that it may well be questioned, whether the mental and moral status of our population has not on the average been lowered, and whether the evil has not overbalanced

> the good. Compared with our wondrous progress in physical science and its practical applications, our system of government, of administering justice, of national education, and our whole social and moral organization, remains in a state of barbarism.

In a long note to defend such an offensive term, Wallace expands on the fate of paupers and criminals in England, and gives examples of the unjust legal system and property laws which exacerbate and perpetuate injustice. (The experience of his early years as a surveyor, working on enclosures, never left him.) 'The wealth and knowledge and culture of *the few* does not constitute civilisation, because it is achieved by means of a mass of human misery and crime absolutely greater than has ever existed before. They create and maintain in life-long labour an ever-increasing army, whose lot is the more hard to bear, by contrast with the pleasures, the comforts, and the luxury which they see everywhere around them, but which they can never hope to enjoy; and who, in this respect, are worse off than the savage in the midst of his tribe.' Until the sympathetic feelings and moral faculties of our nature are more thoroughly trained and developed, and allowed to influence government, commerce and society, 'we shall never, as regards the whole community, attain to any real or important superiority over the better class of savages'. Bringing his argument, and his book, to an emphatic conclusion, Wallace summed up triumphantly: 'This is the lesson I have been taught by my observations of uncivilized man. I now bid my readers – Farewell!'

Wallace was not content to restrict his field of study to the theoretical, or to orang-utans and birds of paradise. For him, the emergence of a new explanation of the origins and development of species had powerful and immediate implications for human behaviour, and for the comparative happiness or misery of human beings. Unlike Darwin, but like his fellow naturalists such as Bates and Spruce, he lived for a long time at the far edges of 'civilization', even though the tentacles of British and Portuguese and Dutch trade were pushing back the frontiers with devastating speed. Certainly in some of the locations he lived in, for all his whiteness and strangeness, he became part of the rhythm of the life around him. As he wrote to his brother-in-law, George Silk, 'I am convinced no man can be a good ethnologist who does not travel, and not *travel* merely, but reside, as I do, months and years with each race'. His sense of that necessary isolation surfaces at key points in his journals and in his book, as here, for example: 'The next day our schooner left for the more eastern islands, and I found myself fairly established as the only European inhabitant of the vast island of New Guinea.' In the Aru Islands, where again he was the only European, he was able to communicate with the islanders in Malay, though their knowledge of the language was less than his. Interestingly, he described the Wanumbai people as 'perfect savages' – he saw no sign of any religion, by which he means Christianity; though some of their customs, such as burying the dead, showed the influence of their long association with Muhammedan traders. These are the people, savage, neither civilised nor Christian, whom he compares so favourably with the barbarism of the English. They were not the idealised noble savage familiar from Rousseau, but people among whom he had lived, resided, for months at a time. The further east he travelled, the more remote the island and the people, the closer he felt he came to an understanding of the origins and nature of man.

What is most remarkable about Wallace is the alternative vision of man, and of the world, which emerges from his experience and writings. Independently, he conceived his theory of natural selection in parallel with Darwin; he was content that Darwin should have the priority, and relieved that it was Darwin who undertook the job of supporting the theory with the great mass of facts and arguments which sustained *The Origin of Species*. Although he eventually parted intellectual company with him in certain areas, he devoted a great deal of his later life to promoting Darwin's views. Yet the social application and potential of his theory emerges, in his voice, quite differently from the mechanistic, power-driven models of some other Darwinians. It was possible, or even temptingly and reductively simple, to harness Darwinian theory to a social and world view which suited the Victorian ruling class perfectly, endorsing capitalism, industrialisation, imperialism, élitism – the survival of the fittest becomes the survival of the strongest, or, in other terms, market forces. The North dominates the South, the capitalists dominate the workforce, the city dominates the country, civilisation eliminates the savage, man destroys nature. While convinced that the Darwinian theory – his theory – explained the development of man, Wallace strove to present an alternative philosophy.

Some of his later enthusiasms and experiments sound odd, even wild. He was interested, as were so many of his contemporaries, in spiritualism, because he believed instinctively in a spiritual dimension, although he firmly rejected the orthodox doctrines of Christianity and had little time for missionaries, except for what they offered in terms of education and medical improvements. But his was, he argued, a rational interest, in that he was willing to believe something possible until it was proved false by experiment or observation. He campaigned against compulsory vaccination, because he considered it was being argued for on the basis of false statistics; and anyway, he did not favour the principle of achieving health through disease. He got into awkward spots, such as his rash acceptance of a bet of £500 to prove that the world was round: he set up an experiment on the Old Bedford Level at Welney, and predictably enough demonstrated his point to the satisfaction of the impartial referees; but his eccentric flat-earther of an opponent cried 'foul', demanded his money back, and pursued Wallace in a series of court cases which brought him unfortunate publicity. He campaigned passionately for land reform, believing that the ownership of land was one of the greatest social evils in England. He was, in fact, an uncompromising socialist. All of this contrived, for a time, to obscure his scientific achievements. It did not help, either, that he became desperately short of money, taking on a number of time-consuming jobs to try and provide for his family, and pursuing, mostly unsuccessfully, various official posts, such as superintendent of Epping Forest.

In these difficult circumstances, Darwin proved a true friend. Hearing from Arabella Buckley, Lyell's former secretary, how hard up Wallace was, Darwin began to organise a petition to Gladstone to grant Wallace a civil list pension. Hooker was reluctant to support him, arguing that Wallace had 'lost caste terribly'; besides, he was not 'in absolute poverty, he just could not find employment'. This was Victorian survival of the fittest in its true colours – charity to the absolutely indigent, if worthy; otherwise, it was a question of pulling yourself up by your own bootstraps. For Hooker, perhaps, Wallace was reverting to his social origins: a labourer who had done well, but who did not belong to the Hooker/Darwin/Lyell world of influence and dynastic marriages. But Wallace's next book, *Island Life,* published in 1880, turned the tide – it was, ironically or fortuitously, dedicated to Hooker, and was widely praised. Huxley supported Darwin; Hooker changed his mind; the president of the Royal Society drafted the petition; even the Duke of Argyll joined in. Gladstone made the recommendation, and in January 1881 wrote to Wallace to tell him that he was to receive a civil list pension of £200, backdated for six months. (A year later, Wallace was one of the pallbearers at Darwin's funeral.) Wallace, by sheer persistence and longevity, emerged as a Grand Old Man and concluded triumphantly with the full range of institutional honours: honorary degrees, prestigious medals – including the Royal Geographical Society's gold medal in 1892. In 1908, the fiftieth anniversary of the Linnean

Society's publication of his Ternate paper, Wallace was awarded the Copley Medal, and, the ultimate accolade, the Order of Merit. He was eighty-five.

Alongside his achievements as an enterprising traveller, a naturalist and a scientific theorist, Wallace thought his way to an independent and often difficult position as a humanitarian. It was difficult, because his beliefs about the spiritual dimension seemed to clash with his otherwise rigorously orthodox views on natural selection. He was one of the first to try to bring the new scientific concept of man into line, not with orthodox Christianity, but with a humanist vision of perfectibility, a perfectibility based on a moral sense far removed from that barbaric struggle he alludes to at the close of *The Malay Archipelago*. This endeavour did not spring from any revisionist process, but was an integral part of his thinking from the first. In 1864, two years after his return to England, he gave a paper to the Anthropological Society of London, 'The Origin of Human Races and the Antiquity of Man deduced from the Theory of "Natural Selection"'. This was one of the first extended attempts at examining natural selection in terms of human races, and Wallace was forced to defend his argument vigorously. His final paragraph, with its poetic dimension and glance at Coleridge, came in for hostile comment.

> In concluding this brief sketch of a great subject, I would point out its bearing upon the future of the human race. If my conclusions are just, it must inevitably follow that the higher – the more intellectual and moral – must displace the lower and more degraded races; and the power of 'natural selection', still acting on his mental organization, must ever lead to the more perfect adaptation of man's higher faculties to the conditions of surrounding nature, and to the exigencies of the social state. While his external form will probably ever remain unchanged, except in the development of that perfect beauty which results from a healthy and well organized body, refined and ennobled by the highest intellectual faculties and sympathetic emotions, his mental constitution may continue to advance and improve till the world is again inhabited by a single homogeneous race, no individual of which will be inferior to the noblest specimens of existing humanity.

Individuals would work out their own happiness in relation to their fellows; perfect freedom of action would be maintained, since well-balanced moral faculties would never permit anyone to transgress on the equal freedom of others: Wallace was drawing on his experience of village communities in the Aru Islands. He believed that physical characteristics were relatively unimportant: it was man's finer intellect which separated human beings from the animals, which allowed man to adapt to the environment, which fostered improvement, and which prompted what he described elsewhere as an 'elastic capacity for co-ordination'. In this utopian vision, 'the passions and animal propensities will be restrained within those limits which most conduce to happiness; and mankind will have at length discovered that it was only required of them to develop the capacities of their higher nature, in order to convert this earth, which had so long been the theatre of their unbridled passions, and the scene of unimaginable misery, into as bright a paradise as ever haunted the dreams of seer or poet'.

Characteristically, Wallace holds back his dystopia of unbridled passions and unimaginable misery, which he saw all around him in mid-Victorian London and which must have seemed infinitely more painful after eight years of travel in paradise, to the very last moment, before shifting into his visionary evocation of Xanadu. Defending his image, he contented himself by commenting: 'I do not think myself that the concluding part of the paper is more poetical than true.' His experience of living among 'savage' peoples convinced him that uncivilised man did have an innate moral sense, which could function effectively at a local, village, egalitarian level, and which was superior, or at least not inferior, to the highly questionable moral system which prevailed in Europe.

7 *The Savage Ape*

THE IDEA OF MAN AS ESSENTIALLY an animal, rather than a being of another, or 'higher', kind, surfaces in so many areas of European culture that it must spring from deeply rooted fears and instincts. The idea is 'primitive', a connection which might more comfortably be suppressed in an advanced industrial society but which scientific observation dragged out into the open. It is central to one of the key Western myths, in the shape of the serpent tempting Eve, a powerful iconographical feature which recurs with unsurprising frequency in the narratives of most travellers to the tropical swamps and forests. In the eighteenth century it features in the rising preoccupation with the Wild Boy, or Wild Girl, the children reportedly brought up in a state of nature by wild animals, such as the Wild Boy of Aveyron, Truffaut's *L'Enfant Sauvage*. Linnaeus catalogued *Homo ferus* as a separate, though human, species, four-footed, mute and hairy. Was man an animal waiting for civilisation to happen, or something else, a special creation made in God's image? Linnaeus had no opportunity to inspect any of the large species which he designated as primates, dead or alive, and his categories of 'man-like apes' contain a strong element of speculation and fiction; Buffon had much better evidence, including some live specimens. In their attempts to piece together a systematic pattern they provided excellent material for the speculative philosopher, Lord Monboddo, a committed evolutionist and a serious anthropologist who had cross-questioned the wild French girl of Champagne. In his *Essay on the Origin and Progress of Language* he wrote:

> Further, not only solitary savages, but a whole nation, if I may call them so, have been found without the use of speech. This is the case of the Ouran Outangs that are found in the kingdom of Angola in Africa, and in several parts of Asia. They are exactly of the human form; walking erect, not upon all-four, like the savages that have been found in Europe; they use sticks for weapons; they live in society; they make huts of branches of trees, and they carry off negroe girls, whom they make slaves of.

4. The Anthropomorpha of Linnaeus.

> . . . These facts are related of them by Mons. Buffon in his natural history; and I was further told by a gentleman who had been in Angola, that there were some of them seven feet high, and that the negroes were extremely afraid of them; for when they did any mischief to the Ouran Outangs, they were sure to be heartily cudgelled when they were catched.

Of the three key indicators – walking upright, communal life, and speech – these creatures of 'fact' possessed two. Monboddo quotes another traveller's tale about 'our species', a variety of men with tails like cats, who navigated in canoes, and who attempted to trade parrots for iron: Monboddo thought that, on the balance of probability, this tailed tribe had not yet invented the art of language, and were thus in the first stage of human progression.

Monboddo, like Rousseau, argued that man and the orang-utan come within the same species. The ape has a long history in European art and literature, with the ape usually employed as a parody of man; and the theme received a powerful boost with Edward Tyson's work of 1699, 'Orang-outang, sive Homo sylvestris; or the Anatomy of a Pygmie compared with that of a Monkey, an Ape, and a Man'. Huxley described this as the 'first account of a man-like ape which has any pretensions to scientific accuracy and completeness'. Swift's Yahoos owe something to Tyson's orangs, and other European writers to explore the theme include Restif de la Bretonne, E.T.A. Hoffmann and Edgar Allan Poe (in his 1841 story, 'The Murders in the Rue Morgue', it was the orang-utan who did it). When Mephistopheles takes Faust to visit the witch's kitchen in search of a rejuvenating liquor, Goethe creates a disturbing and grotesque scene with two apes who stir the pot, making thin soup for beggars, while their young play with a big ball. The ambivalence of the image is highlighted by the contrasting responses: Faust finds the apes and their conversation 'As gross as anything I ever saw', while Mephistopheles says, 'Now for me, a chat like this is just what I enjoy.' Thomas Love Peacock's 1817 satire *Melincourt*, which glances at Monboddo's theories, pushes into the political arena. The central character Forester adopts an orang, and tries unsuccessfully to teach it language. Although it remains dumb, it does learn to play the flute, obtains a baronetcy, and, as Sir Oran, enters the House of Commons as representative of a rotten borough.

Many of these literary and imaginative explorations of the ape were satirical or philosophical rather than strictly scientific, but they were informed by scientific enquiry. The Shelleys, for example, were friends of the evolutionist Sir William Lawrence, at that time their physician, who opened up questions concerning generation and the nature of life. (Huxley commented in 1894 that Lawrence had been well-nigh ostracised for his book *Lectures on Physiology, Zoology and the Natural History of Man*, 'which now might be read in a Sunday-school without surprising anybody'; ostracism seems a mild term – Lawrence was suspended by the Royal College of Surgeons, which refused to reinstate him until he withdrew the book. This resulted in several unauthorised editions, because of a wonderfully obtuse ruling by the Lord Chancellor that an author was not protected by copyright where a work was blasphemous, seditious or immoral.) The idea of creating a new species is one of the mainsprings of Mary Shelley's *Frankenstein*, though the monster which Victor Frankenstein brings to life is more human than ape. Significantly, although Frankenstein proclaims that he has 'pursued nature to her hiding places', his science is described exclusively in negative terms. 'Who shall conceive the horrors of my secret toil, as I dabbled among the unhallowed damps of the grave, or tortured the living animal to animate the lifeless clay?' The charnelhouse, the dissecting room and the slaughterhouse provide his context. The Creature who emerges, and from whom Frankenstein flees, resembles a primitive man, though eight feet tall; he has no language, lives off berries and roots, and receives his education by observing what appears to be an ideal family through a chink in a cottage wall. But he is rejected both by them and by his creator, and finally disappears on an ice-raft into the darkness of the Arctic night, not permitted a place in human society by a competitive, self-regarding aristocracy. The monster is not specifically described as an ape: Frankenstein calls him monster, fiend, devil, vile insect; he is ape-like in that he has the figure of a man, but is denied man's attributes. He appears monstrous, and is rejected as such. In the first stage adaptation in 1823, he was even denied the power of speech. He should be Adam, but (having read *Paradise Lost*) knows himself to be a fallen angel. Frankenstein rejects him, as he has rejected him from the moment of his creation: 'There can be no community between you and me; we are enemies.' Although Frankenstein consents under pressure to make a female companion of the same species, he reneges on his promise; and as the story concludes, he has become animal-like himself, pursuing his monster over the snowy wastes, a monster who by this time has learned to imitate the worst traits of humanity. To complete the pattern, the overarching narrator is a scientific traveller, aiming to discover a sea route through the Arctic Ocean. 'I shall kill no albatross, therefore do not be alarmed for my safety,' he writes to his sister.

Frankenstein, among other meanings, presents the exploration of one species' relationship with another; and the ape – chimpanzee, orang-utan, gorilla – appealed strongly to the popular imagination, as much as to the scientist, as the species which most resembled man. In 1859, the year when *The Origin of Species* was published, Paul du Chaillu could claim to be the first white man to have seen a live gorilla in its natural habitat, and Alfred Wallace, four years before, was the first English naturalist to make a systematic study of the orang in the wild. The more man-like the apes, the more they disturbed: footprints in the sand which could not be washed away.

The ape, then, had a powerful hold on the Western imagination long before the link between man and monkey was celebrated in the polarised debate in 1860 between science, with Huxley as its champion, and the Church, in the person of Samuel Wilberforce, the Bishop of Oxford. In the late twentieth century, zoologists go to the rain forest and squat in chimpanzees' nests, and the events are photographed or filmed. In the nineteenth, the animals were brought back to Europe, to be displayed in man-made tropical environments if they were lucky, or in the more sensationalist environment of a travelling menagerie. The great apes came relatively late on the European scene: the first orang-utan reached France in 1720, in time to be classified by Buffon. The London

Zoo only acquired an orang in 1837, put her in the well-heated giraffe house, and named her Jenny: she was the first to live through a winter, and go on show. A successor, also named Jenny, was inspected by Queen Victoria, who commented: 'The Orang outang is too wonderful preparing and drinking his tea, doing everything by word of command. He is frightful and painfully and disagreeably human.'

Charles Waterton, traveller and taxidermist, who turned his Yorkshire estate into a game park, had a stuffed baboon strung from his ceiling at Walton Hall among his many trophies. Waterton was a naturalist unlike any other of his time: wandering the forests of South America barefoot in the first decades of the nineteenth century, wading through the rivers, he liked to observe nature at close quarters, and was quite content to let insects bite him, on the grounds that this would probably not make you ill for very long. He tended to grapple with his quarry hand to hand: 'I have attacked and slain a modern Python, and rode on the back of a cayman close to the water's edge; a very different situation from that of a Hyde-park dandy on his Sunday prancer before the ladies. Alone and barefoot I have pulled poisonous snakes out of their lurking-places; climbed up trees to peep into holes for bats and vampires. . . .' It was all true: when asked how he kept his seat on the cayman's back, he put it down to his years of experience with Lord Darlington's foxhounds. He was also something of a controversialist, quarrelling bitterly with Audubon, and taking issue with the new science of taxonomy. In 1857 he jumped on the monkey bandwagon, and offered a plain man's guide for the young naturalist as an antidote to the zoologists who had 'fabricated systems so abstruse, so complicated, and so mystified'. Waterton, as it happened, had rather limited experience of monkeys, apart from the apes on Gibraltar and a few species in South America. However, he had had the good fortune to have 'made the acquaintance in England' of three species of ape from the warm regions of the tropics, and he wove them into his entertaining but intermittently inaccurate essay, 'A New History of the Monkey Family'. Waterton cultivated his connections with the Wombwell family, who ran a menagerie, and acquired a chimpanzee, who died of bronchial trouble in Scarborough. Waterton even suggested that it should be exhibited dead at Huddersfield. The 'frosty state of the weather was all in its favour', and when it eventually arrived, he spent seven weeks restoring its form and features, through his special taxidermist's technique, sitting it on a cocoa-nut which he had brought from Guiana in 1817.

25. A gorilla, pickled in rum. (Photograph, 1858, London Zoo)

Another important contact for Waterton was Mitchell, the secretary of the London Zoo, who received an orang-utan from Borneo in 1851. Waterton was given permission to enter its cage:

As I approached the orang-utan, he met me about half-way, and

> we soon entered into an examination of each other's persons. Nothing struck me more forcibly than the uncommon softness of the inside of his hands. Those of a delicate lady could not have shewn a finer texture. He took hold of my wrist and fingered the blue veins therein contained; whilst I myself was lost in admiration at the protuberance of his enormous mouth. He most obligingly let me open it, and thus, I had the opportunity of examining his two fine rows of teeth.
>
> We then placed our hands around each other's necks; and we kept them there awhile.

Waterton was not too absorbed to notice the effect on the spectators outside the cage, who seemed 'wonderfully amused at the solemn farce before them'. The orang only survived another nine months, and Waterton was bitterly disappointed to miss out on the corpse, through Mitchell being away at the critical moment.

His third example was identified as a chimpanzee, though it was almost certainly a gorilla, from the River Congo. This, too, formed part of the Wombwell menagerie, and was consigned to a particularly miserable existence, being dressed all the time in human clothes, and living in an attic room with a female keeper. Waterton was fascinated, and desperately sorry for Jenny: 'Her skin is as black as a sloe in the hedge, whilst her fur appears curly and brown. Her eyes are beautiful; but there is no white in them; and her ears are as small in proportion as those of a negress.' When he saw her for the last time he projected his sympathy into her thoughts: 'The little room is far too hot; the clothes which they force me to wear are quite insupportable, whilst the food which they give me is not like that upon which I used to feed, when I was healthy and free in my own native woods.' When Jenny died, her keeper wrapped her up in a sheet, popped her in a trunk, and forwarded her to Waterton. She was as badly treated in death as she had been alive. He decided to turn her into one of his taxidermist's freaks, equipped her with a pair of donkey's ears, and christened the composite Martin Luther.

The largest ape, the gorilla, was the last on the scene. Two skulls reached London in 1846, from an American missionary, Dr Thomas Savage, to be followed by a skeleton in 1851 and a specimen pickled in alcohol in 1858. These were all measured and described by Owen, who had dissected his first ape, a young orang, in 1830. Richard Owen, Hunterian Professor of the Royal College of Surgeons and, from 1856, in charge of the natural history section of the British Museum, was the leading comparative anatomist of his day. In 1857 he had thrown down a marker by stating that man was different from other mammals, because of a distinctive brain. In philosophical terms, man had a moral dimension and a rational capacity which the apes lacked. In anatomical terms, this could be demonstrated by man's possession of the 'hippocampus minor', a structure lacking in apes. Huxley responded in 1858 in a lecture, 'The Distinctive Characters of Man': 'Now I am quite sure that if we had these three creatures [humans, gorillas and baboons] fossilized or preserved in spirits for comparison and were quite unprejudiced judges we should at once admit that there is very little greater interval as animals between the Gorilla & the Man than exists between the Gorilla and the Cynocephalus [the baboon].' Huxley's claim was uncompromising and stark: 'to the very root & foundation of his nature man is one with the rest of the organic world'.

When the French-American explorer Paul du Chaillu came back from Gabon in equatorial Africa with an impressive array of skulls, skeletons and stuffed specimens, backed by vivid stories of his face to face encounters with the creatures, he was soon invited to exhibit them in New York, while *Harper's Magazine* commissioned him to write up his travels. The Royal Geographic Society decided the opportunity was too good to miss, and suggested he brought the exhibition to London. A room was made available, and he was invited to present a report at the Society's annual meeting.

Du Chaillu was something of an enigma, a perplexing amalgam of hunter, naturalist, explorer, showman and journalist. He was a short, pugnacious man, and spoke English with a heavy French accent. His scientific credentials were partly verified by his contacts with American professors and learned societies. Among the English scientific community, he aligned himself with Owen – he sold his collection to the British Museum – which was enough in itself to arouse Huxley's scepticism. His book, *Explorations and Adventures in Equatorial Africa*, was published in London by Murray in May 1861: the second edition, two months later, added a rough chronology, since 'discrepancies' in the dates had been pointed out to the author. The book reflected du Chaillu's double role, as naturalist and entertainer: the naturalist collates the information, and offers a systematic account of the various categories or species of apes he has encountered; the entertainer presents the intrepid explorer and his sensational adventures: the civilised European among the savages and cannibals, the Christian among the pagans, the lone hunter confronting serpent and wild beast in swamp and mountain forest. Du Chaillu travelled, he announced, about 8,000 miles, always on foot and unaccompanied by other white men; he shot and stuffed and brought home over 2,000 birds, including 60 new species, and 200 quadrupeds, with 80 skeletons, of which no fewer than 20 were hitherto 'unknown to science'. For all this massive accumulation of specimens, however, the focus of the book is undoubtedly the gorilla; one crouches embossed on the book's binding; another, adapted from Geoffroy St Hilaire's illustration, unfolds as a frontispiece to stand, proud and menacing, facing the title page. The first reference in the Preface is to 'that monstrous and ferocious ape, the gorilla'; and it is the hunt for the gorilla's habitat, in the hills inland from the Gaboon river, which provides the structure for the slightly confusing pattern of travelling: significantly, the white explorer finds them in the country of the Fan, by reputation the most ferocious of cannibal tribes.

Just before his first encounter, du Chaillu's party had feasted off a great snake, which he felt unable to stomach (though he later overcame his scruples about eating monkey). Instead, he picked some sugar-cane to chew, and discovered that the gorillas had been there first.

26. Paul du Chaillu and party surprised by an enormous black snake in a mangrove-swamp. 'I came very near getting a mud-bath myself.' (*Explorations and Adventures in Equatorial Africa*)

> We followed these traces, and presently came to the footprints of the so-long-desired animal. It was the first time I had ever seen these footprints, and my sensations were indescribable. Here was I now, it seemed, on the point of meeting face to face that monster of whose ferocity, strength, and cunning the natives had told me so much; an animal scarce known to the civilized world, and which no white man before had hunted. My heart beat till I feared its loud pulsations would alarm the gorilla, and my feelings were really excited to a painful degree.

It is the Robinson Crusoe moment. Du Chaillu attempts to build up the excitement: they were 'armed to the teeth'; the male gorilla was 'literally the king of the African forest; 'we were about to pit ourselves against an animal which even the leopard of these mountains fears'. He also adds a few 'native stories' about women being carried off into the woods by gorillas. But when he finally catches up with the animals, and is startled 'by a strange, discordant, half human, devilish' cry, he sees four young gorillas running towards the deep forests. The men fire, but miss their targets: 'The alert beasts made good their escape.' The face to face encounter is postponed.

Du Chaillu at once raises the question of the human appearance of the gorillas. As they ran, on their hind legs, they looked 'fearfully like hairy men; their heads down, their bodies inclined forward, their whole appearance like men running for their lives'. It was not surprising that the natives held strange superstitions about these 'wild men of the woods'.

These anthropomorphic feelings did not deter du Chaillu. He set off with the Fan on a gorilla hunt, and was soon rewarded by finding himself confronting an immense male: 'Nearly six feet high [he proved four inches shorter], with immense body, huge chest, and great muscular arms, with fiercely glaring large deep gray eyes, and a hellish expression of face, which seemed to me like some nightmare vision: thus stood before us this king of the African forest.' The gorilla beat his breast, and vented roar after roar.

> His eyes began to flash fiercer fire as we stood motionless on the defensive, and the crest of short hair which stands on his forehead began to twitch rapidly up and down, while his powerful fangs were shown as he again sent forth a thunderous roar. And now truly he reminded me of nothing but some hellish dream creature – a being of that hideous order, half-man half-beast, which we find pictured by old artists in some representations of the infernal regions. He advanced a few steps – then stopped to utter that hideous roar again – advanced again, and finally

27. Hunter killed by a gorilla. Another Paul du Chaillu incident. 'The stock was broken, and the barrel was bent and flattened. It bore clearly the marks of the gorilla's teeth.'

> stopped when at a distance of about six yards from us. And here, just as he began another of his roars, beating his breast in rage, we fired, and killed him.

Du Chaillu skirts round the nature of the gorilla and its relationship with man, in all his descriptions. Each time he kills one, he raises the same question. 'There is enough likeness to humanity in this beast to make a dead one an awful sight,' he comments on a later kill. 'It was as though I had killed some monstrous creation, which yet had something of humanity in it. Well as I knew that this was an error, I could not help the feeling.' The same ambivalence permeates his descriptions of the mountain Fan, whose energy and resilience he admired; his descriptions of elaborate precautions against eating human flesh, or even eating from a cooking-pot which might have contained human flesh, seem designed to reassure his essential, European, Christian difference even as his narrative indicates how closely he integrated himself. He takes his reader with him into nightmare, into the infernal regions, slaying monsters who are disturbingly made almost in his own image; and then he brings back the skulls and the skeletons as scientific proof.

Du Chaillu seems never to have investigated the apes he hunted in their natural habitat; but he did acquire a young ape, of a species he identified as 'nshiego mbouve', a kind of chimpanzee, whose mother had been shot. 'His little eyes became very sad, and he broke out in a long plaintive wail.' Du Chaillu's heart ached to see the little creature, who looked quite forlorn, 'as though he really felt his forsaken lot'. The baby's face was pure white, as white as a white child's.

> While I stood there, up came two of my hunters and began to laugh at me. 'Look, Chelly!' said they, calling me by the name I was known by among them, 'look at your friend. Every time we kill gorilla, you tell us, "Look at your black friends!" Now, you see, look at your white friend!' Then came a roar at what they thought a tremendously good joke.
>
> 'Look! he got straight hair, all same as you. See white face of your cousin from the bush! He is nearer to you than gorilla is to us.'
>
> And another roar.
>
> 'Gorilla no got woolly hair like we. This one straight hair, like you.'
>
> 'Yes,' said I; 'but when he gets old his face is black; and do not you see his nose how flat it is, like yours?'

Du Chaillu tamed the ape in three days, and named him Tommy. He soon showed signs of original sin, and stole plantains or fish: du Chaillu flogged him, and claimed to have 'brought him to a conviction that it was *wrong* to steal'; but he could never resist the temptation. Du Chaillu made him a little pillow to sleep on, which he became as attached to as a child to a comfort blanket; and he always joined him at mealtimes, acquiring a taste for coffee, 'Scotch ale', and even brandy. But plans to take this 'wonderful little creature', who was apparently adapting to civilised life, back to America came to nothing, when he sickened and died. 'Alas! poor Tommy!'

Later in his book, du Chaillu gives a more objective account of the gorilla and the other African apes, dispelling, though taking care to repeat, the more lurid stories about the gorilla's habits which he had reported earlier. He included, for example, a tale about a woman who had been carried off by a gorilla, though his version, 'She related that the gorilla had forced her to submit to his desire', was toned down at Murray's insistence to 'She related that the gorilla had misused her.' He still emphasised, as the first white man who could speak from personal knowledge, the 'horror of its appearance, the ferocity of its attack', and the 'impish malignity of its nature'. He also admitted that it was that 'lurking reminiscence of humanity' which was one of the chief ingredients of his hunter's excitement as he confronted it. Nevertheless, in common with Owen, and perhaps advised by him, du Chaillu was swift to reiterate the great dissimilarity between the bony frame of man and that of the gorilla (though there was also 'an awful likeness'): the vertebral column was essentially curved, and the brain was structurally different and distinctive, as well as larger. *Homo sapiens* could relax. Besides, he concluded, he had made an extensive search for an 'intermediate race' or rather several intermediate races or links between the natives and the gorilla. 'But I have searched in vain,' he reported, as though disposing of the matter. 'I found not a single being, young or old, who could show an intermediate link between man and the gorilla, which would certainly be found if man had come from the ape.' Du Chaillu aligned himself with the anti-evolutionists, and appeared, plus stuffed gorilla, at the Metropolitan Tabernacle in Newington, with the archaeologist 'Nineveh' Layard in the chair, at one of the Baptist preacher Charles Spurgeon's most successful meetings. The Tabernacle had been completed in March 1861, and held 6,000. The occasion generated a rash of cartoons, including the inevitable 'A gorilla lecturing on Mr Spurgeon'.

Du Chaillu certainly caused a stir, both with his lectures and his book. Heinrich Barth was scathing: he accused du Chaillu of inventing his itinerary; he had taken no scientific instruments with him, so how could he know where he was? Charles Waterton challenged him to produce some evidence – 'I am quite ready for him' – called him a disgrace to zoology, and suggested he had never seen a live gorilla. Richard Burton, in contrast, spoke up for him at the Ethnological Society, where du Chaillu lost his temper with a questioner, spat in his face, and thumped him. Doubts remained, not least in Huxley's mind. Du Chaillu's accounts bore a remarkable similarity to those of Dr Savage, published in the *Boston Journal of Natural History* in 1847, supplemented by a further paper by another American, Ford, given to the Philadelphian Academy of Sciences in 1852. Both men made it clear that they were drawing on accounts by native Africans; their descriptions, for instance of how a gorilla will seize and crush a musket between his teeth, could easily serve as a basis for one of du Chaillu's anecdotes. In his 1863 essay on 'The Natural History of the Man-like Apes' Huxley quoted Ford's description in full, 'for comparison with other narratives'; he goes on to assert:

> If subtraction be made of what was known before, the sum and substance of what M. Du Chaillu has affirmed as a matter of his own observation respecting the Gorilla is, that, in advancing to the attack, the great brute beats his chest with his fists. I confess I see nothing very improbable, or very much worth disputing about, in this statement. . . . If I have abstained from quoting M. Du Chaillu's work, then, it is not because I discern any inherent improbability in his assertions respecting the man-like Apes; nor from any wish to throw suspicion on his veracity; but because, in my opinion, so long as his narrative remains in its present state of unexplained and apparently inexplicable confusion, it has no claim to original authority respecting any subject whatsoever. It may be truth, but it is not evidence.

This is a neat example of how to discredit someone without quite calling him a liar. In contrast, Huxley gives full and generous recognition to Wallace, whose papers on the orang-utan he draws on: 'Once in a generation a Wallace may be found physically, mentally, and morally qualified to wander unscathed through the tropical wilds of America and Asia; to form magnificent collections as he wanders; and withal to think out sagaciously the conclusions suggested by his collections.' A man might well be excused, given the dangers from disease alone, if he contented himself with 'stimulating the industry of the better seasoned natives' and collecting and collating their more or less mythical reports. For Huxley, du Chaillu was a hunter and journalist masquerading as a scientific naturalist.

Du Chaillu, though he almost certainly fabricated or elaborated some of his encounters, was an unusually resilient and risk-taking explorer, with a strong belief in the protection of quinine; many of the details of his account of the Fan, about which many of the London scientists were equally sceptical, were later corroborated by other travellers and ethnologists. He made a second trip in 1863, after taking instruction in how to use a camera and other scientific instruments; this time he managed to obtain a live gorilla, though it died on the voyage. When he returned to London Henry Walter Bates offered to help him with his notes: recognition from one traveller to another, and advice on how to present findings more objectively. To a great extent, du Chaillu was vindicated.

The separateness of man and ape, not only in the here and now but in the past, was a fiercely disputed frontier. Was man, *Homo sapiens*, a different order of being, of creation, with a different kind of brain and a spiritual dimension; or an evolved form, evolved, in its most extreme and radical expression, by chance and random variations which happened to survive and breed successfully? Du Chaillu confronting the gorilla in equatorial Africa, standing eight yards away with levelled rifle; Queen Victoria quizzing Jenny the orang-utan at the London Zoological Gardens – were they gazing at something frighteningly or comically similar but essentially other; or were they confronting as if in a mirror their own image? Towards the end of the century a remarkable naturalist called Richard Garner reversed the process by taking up his quarters in a collapsible steel-wire cage in the African forest 200 miles up the Ogowe river. It was painted a dingy green, and protected with bamboo leaves and canvas curtains. Garner took up residence in this ingenious hide, named Fort Gorilla, so as to study chimpanzees and gorillas in relative security, and 'to live among them' for over three months. He was accompanied by a young chimpanzee called Moses, who slept in a hammock next to his camp-bed, and perched on his shoulder when he ventured out into the bush. Garner chattered to Moses, whom he had found in a papyrus swamp by the Ogowe, in his own chimpanzee language, and taught him to pronounce 'feu'; needing a witness to his own signature on a legal document, he guided his hand in making a cross, and annotated it: Moses Ntyigo His Mark – the ultimate elevation of Victorian ape to human status. Garner remained convinced that, in time, he would have taught Moses to speak.

Within Darwin's huge framework, Huxley and Owen squared up to each other on a key question of anatomy. Owen declared that a gorilla's brain was closer to a lemur's than to a man's, for the gorilla lacked a cerebrum and hippocampus; Huxley argued that, 'in the important matter of cranial capacity, Men differ more widely from one another than they do from the Apes'. It was a dispute which was as much philosophical, theological and psychological, as scientific. Huxley and Owen pitched into each other in the columns of the *Athenaeum*. The debate was also carried on at a popular level. In the *Punch* of 15 May 1861 appeared a lampoon under a picture of a gorilla, and the Abolitionist caption: 'Am I a Man and a Brother?':

> Next HUXLEY replies
> That OWEN he lies
> And garbles his Latin quotation;
> That his facts are not new,
> His mistakes not a few,
> Detrimental to his reputation.
> 'To twice slay the slain,'

By dint of the Brain
(Thus HUXLEY concludes his review)
Is but labour in vain,
Unproductive of gain,
And so I shall bid you 'Adieu!'

The verses were signed 'Gorilla', and were written by a scientist friend of Owen's, Sir Philip Egerton.

Huxley, seconded by Hooker and Lubbock, had tackled the conservative establishment head on when he rose to respond to Bishop Wilberforce's jibe at the meeting of the British Association in Oxford in 1860: the Bishop ill-advisedly enquired whether the apes were on his grandfather's or grandmother's side of the family. Huxley must have known that sooner or later there would be an opportunity for open and public conflict, but he had not anticipated anything quite so tailor-made. He turned to his neighbour and commented, 'The Lord has delivered him into my hands', then waited calmly until he was asked to reply. Huxley's version of a much-discussed and variously reported exchange ran: 'If then, said I, the question is put to me would I rather have a miserable ape for a grandfather or a man highly endowed by nature and possessed of great means of influence & yet who employs these faculties & that influence for the mere purpose of introducing ridicule into a grave scientific discussion, I unhesitatingly affirm my preference for the ape.' Hooker, also present, took up the battle, and, in his own words to Darwin, 'smashed him amid rounds of applause. I hit him in the wind at the first shot in ten words taken from his own ugly mouth; and then proceeded to demonstrate in as few more: (1) that he could never have read your book, and (2) that he was absolutely ignorant of the rudiments of Bot. Science.' Darwin, Hooker concluded triumphantly, was master of the field.

In 1860 Huxley took 'The Relation of Man to the Lower Animals' as his subject for six lectures he gave to working men; as he commented to his wife, his men stuck by him wonderfully: 'By next Friday evening they will all be convinced that they are monkeys.' In January 1862, he was invited to lecture on the same subject to the Philosophical Institute of Edinburgh. These lectures formed the basis of essays published in 1863 as *Man's Place in Nature*. Huxley's clarity and incision remains as fresh today as when he educated the intelligent working people of London: 'The question of questions for mankind – the problem which underlies all others, and is more deeply interesting than any other – is the ascertainment of the place which Man occupies in nature and of his relations to the universe of things. Whence our race has come; what are the limits of our power over nature, and of nature's

28. Skeletons of the Gibbon, Orang, Chimpanzee, Gorilla, and Man. 'Photographically reduced from Diagrams of the natural size (except that of the Gibbon, which was twice as large as nature), drawn by Mr Waterhouse Hawkins from specimens in the Museum of the Royal College of Surgeons.' (Huxley, *Man's Place in Nature*)

power over us; to what goal we are tending'. By the end of the essay, Huxley, step by inexorable step, has demonstrated that 'the structural differences which separate Man from the Gorilla and the Chimpanzee are not so great as those which separate the Gorilla from the lower apes'. That is not to say the structural differences are not great and significant; every bone of a gorilla bears marks which distinguish it from the corresponding bone of a man. There is a chasm between the two, no link in the present creation between *Homo* and *Troglodytes*; man is a family apart, but belonging to the same order for which the Linnaean term 'primates' should be retained; and the explanation of man's origins, he concludes, is by gradual modification within the hypothesis put forward by Darwin.

Huxley, for all the lucidity of his scientific reasoning and the vigour of his debating technique, never lost sight of the philosophical dimension. Was man debased by this demonstration of his origins, no more than a poor, bare, forked animal? For Huxley, the answer was a resounding no.

> Our reverence for the nobility of manhood will not be lessened by the knowledge that Man is, in substance and in structure, one with the brutes; for, he alone possesses the marvellous endowment of intelligible and rational speech, whereby, in the secular period of his existence, he has slowly accumulated and organised

> the experience which is almost wholly lost with the cessation of every individual life in other animals; so that, now, he stands raised upon it as on a mountain top, far above the level of his humble fellows, and transfigured from his grosser nature by reflecting, here and there, a ray from the infinite source of truth.

Huxley's language is inflected with the language of the New Testament, and, in a metaphor of progress and ascent, he celebrates the transfiguration of ape into man. Darwin rejoiced when the 'Monkey Book' was published: it was like a postscript to his own *magnum opus*, but with the focus on man, as the title announced so unambiguously. The search for man's origins would now be turned to the fossil record, in an attempt to establish the 'missing links'. The debate was far from over; but Huxley had swept aside myth and prejudice and laid out for all to see the structural affinities between man and ape, adjusting by fact what Linnaeus had attempted by guesswork. Where du Chaillu's gorillas appeared monstrous, even demonized, Huxley's skeletons of gibbon, orang, chimpanzee and gorilla led plainly and unsensationally towards upright man, in a continuous and progressive dance of life.

8 *Natural Perspectives*

WHEN SO MANY OF THE SCIENTIFIC explorers and naturalists sailed under the auspices of the British navy, or with the British government's direct or indirect assistance, it is hardly surprising that women found it hard to break into the field. Amelia Edwards, the Egyptologist, and Margaret Fountaine, the butterfly collector, were just two women who won acceptance slowly in their respective specialities. Only enormous personal wealth and willpower could overcome that set of prejudices, as in the case of the Anglo-Dutch heiress Alexandrine Tinne. In 1862 Tinne, in the company of her mother and aunt, pushed up the Nile with 300 men and 100 camels, and published *Plantae Tinneanae* as a grand gesture to her scientific contribution. Her second expedition, for which she hired two German scientists, was thwarted when fever fatally struck most of her party. Turning her back on Europe, she resolved to be the first woman to cross the Sahara, and planned to strike eastwards from Lake Chad towards the Nile. Neither her wealth nor her fame could save her. She was murdered in the desert by Tuaregs in 1869.

Two very different and equally remarkable women who brought unusual perspectives to their scientific travelling were Mary Kingsley and Marianne North, the one as a collector and writer, the other as a botanical artist.

Mary Kingsley explored in the mangrove swamps and river systems of West Africa, and turned the experience into a remarkable piece of travel writing, *Travels in West Africa*, published in 1897, which she followed with the more ethnographical *West African Studies* two years later. Her father, George Kingsley, Charles Kingsley's brother, was a doctor who spent the last years of his life travelling abroad; only after her parents' death in 1892, it seems probable, did his daughter, his elder child, realise that he had married just a few days before her birth. Father and mother died within three months: Mary had been nursing her paralysed mother, wheeling her round Parker's Piece in a bathchair from her Cambridge home. Released from her role of dutiful daughter, she made a trip to the Canaries, and then chose to steam down the coast of

West Africa on a succession of cargo ships and Portuguese mail-boats, one of the more bizarre and uncomfortable experiences available to the late Victorian tourist. She was thirty-one. According to her Preface to *Travels in West Africa*, she went at the call of Science, and on the indirect prompting of Alfred Wallace: '"Go and learn your tropics", said Science. Where on earth am I to go, I wondered, for tropics are tropics wherever found, and so I got down an atlas and saw that either South America or West Africa must be my destination, for the Malayan region was too far off and too expensive. Then I got Wallace's Geographical Distribution and after reading that master's article on the Ethiopian region I hardened my heart and closed with West Africa.' She also went, in a sense, to complete her father's work, for his studies had concentrated on primitive religions and sacrificial rites: hence her preoccupation with 'fetish', while 'fish', her chosen branch of zoology, provided a sound scientific motive. For the most part, she stayed on board during this first trip, but she sent her visiting card ashore at Cabinda, following it up in person, and billeted herself on Richard Dennett, in charge of the Hatton and Cookson trading station. This visit changed her mind about the Old Coaster 'palm oil ruffians', and she became a fierce champion of trade, and of the morality of the traders: 'of their kindness to me I can never sufficiently speak, for on that voyage I was utterly out of touch with the governmental circles, and utterly dependent on the traders, and the most useful lesson of all the lessons I learned on the West Coast in 1893 was that I could trust them.' Back in Liverpool, Mary Kingsley walked about with a monkey on her shoulder: she had been marked and changed by her first African experience.

She began to carve out a role for herself, as a serious naturalist. She had brought back a small collection of specimens from her first reconnaissance, and she chose to specialise in 'fish and fetish'. She arranged to show her specimens to Dr Gunther at the British Museum, and he in due course supplied her with fifteen gallons of spirits to equip her as a bona fide collector for her second expedition. She also sent her journals to Macmillan's, with a view to having her account of these next travels published. She sailed from Liverpool in December 1894, in the company of Lady Ethel MacDonald, the wife of the consul-general of the Niger Coast Protectorate, which certainly placed her firmly in touch with government circles. She spent a considerable amount of time with the MacDonalds, a fact she tends to gloss over in the *Travels*, though she does record her visit to the great missionary, Mary Slessor, a woman whose unconventional and forthright approach to Africa and Africans was very much in tune with her own.

Some commentators have wondered about the objectivity of Mary Kingsley's narrative. Dates and places do not feature in any systematic way. She describes the dangers and discomforts with a robust directness, and an ironic understatement, which is entertaining and disarming. For example, 'on being such a colossal ass as to come fooling about in mangrove-swamps' (which she frequented in search of fish), she recollected:

> On one occasion, the last, a mighty Silurian, as the *Daily Telegraph* would call him, chose to get his front paws over the stern of my canoe, and endeavoured to improve my acquaintance. I had to retire to the bows, to keep the balance right, and fetch him a clip on the snout with a paddle, when he withdrew, and I paddled into the very middle of the lagoon, hoping the water there was too deep for him or any of his friends to repeat the performance. Presumably it was, for no one did it again. I should think that crocodile was eight feet long; but don't go and say I measured him, or that this is my outside measurement for crocodiles. I have measured them when they have been killed by other people, fifteen, eighteen, and twenty-one feet odd. This was only a pushing young creature who had not learnt manners.

Almost certainly she was not alone on this occasion, although she tends to write, with deliberate irony, in the style of the great white hunter in an eyeball to eyeball brush with nature. But the astonishing thing was that she had gone paddling about in mangrove swamps at all, and that, despite the crocodiles and the mangrove flies and the fearful stench of the black, batter-like slime, she enjoyed the experience. She positively appreciated the strange and unique nature of the terrain, so much so that she could write: 'I believe the great swamp region of the Bight of Biafra is the greatest in the world, and that in its immensity and gloom it has a grandeur equal to that of the Himalayas. I am not saying a beauty; I own I see a great beauty in it sometimes, but it is evidently not of a popular type'. The mangrove swamp and the African forest became Mary Kingsley's chosen habitats; and, out of all the tribes she met, she, perhaps inevitably, selected the Fan, whose reputation for ferocity and cannibalism was proverbial, as her favourite:

> As it is with the forest, so it is with the minds of the natives. Unless you live alone among the natives, you never get to know them; if you do this you gradually get a light into the true state of their mind-forest. At first you see nothing but a confused

> stupidity and crime; but when you get to see – well! as in the other forest – you see things worth seeing.

The expedition which lies at the heart of *Travels in West Africa* was Mary Kingsley's journey down the Ogooué (or Ogowe), in Gabon, and then across land – or swamp – to the Rembwé: an area under French influence, through which Paul du Chaillu had trekked thirty-five years before, and about which he had written so sensationally. There was no map in her book: the particular route she took, as Dea Birkett observes, 'was approximately fifty miles as the crow flies and would take her eight days. It was a route no white person had attempted before, most probably because of its insignificance.' But she did it; and she did it as a white woman, not speaking any relevant language, without the protection of government or missionary status. She had the right to make the most of it: encounters with elephants, hippopotamus, crocodiles, a party of gorillas; sharing a snake for supper with the Fan; falling into a game-pit, when the fullness of her skirt saved her from the impact of nine ebony spikes (if she had followed the advice of people in England and adopted masculine garments, she would have been skewered to the bone). The short expedition, undoubtedly uncomfortable and dangerous, became in the telling surprisingly crammed with startling incidents.

Staying in a Fan village during the trip, she found herself unable to sleep – 'mosquitoes! lice!!' – so she got up and went down to the shore of the lake. She quietly slid a small Fan canoe into the water, helped herself to a paddle from a cluster stuck in the sand, and pushed off:

> It was a wonderfully lovely quiet night with no light save that from the stars. One immense planet shone pre-eminent in the purple sky, throwing a golden path down on to the still waters. Quantities of big fish sprung out of the water, their glistening silver-white scales flashing so that they look like slashing swords. Some bird was making a long, low boom-booming sound away on the forest shore. I paddled leisurely across the lake to the shore on the right, and seeing crawling on the ground some large glow-worms, drove the canoe on to the bank among some hippo grass, and got out to get them.
>
> While engaged on this hunt I felt the earth quiver under my feet, and heard a soft big soughing sound, and looking round saw I had dropped in on a hippo banquet. I made out five of the immense brutes round me, so I softly returned to the canoe and shoved off, stealing along the bank, paddling under water, until I deemed it safe to run out across the lake for my island. I

> reached the other end of it to that on which the village is situated; and finding a miniature rocky bay with a soft patch of sand and no hippo grass, the incidents of the Fan hut suggested the advisability of a bath. Moreover, there was no china collection in that hut, and it would be a long time before I got another chance, so I go ashore again, and, carefully investigating the neighbourhood to make certain there was no human habitation near, I then indulged in a wash in peace. Drying one's self on one's cummerbund is not pure joy, but it can be done when you put your mind to it.

This starlit naked bathe in a lake in the depths of the great forest, in the country of the notorious Fan, suggests Mary Kingsley's total absorption in Africa, as well as her courage and independence. She had come a long way from the stuffy respectability of Cambridge. Significantly, she is careful to maintain the role of the scientific naturalist alongside the more personal experiences, although in a slightly mocking tone. Before leaving the hut, she records that she had been rearranging the luggage, unpacking her bottles of fishes to equalise the weight of the loads; and while she was finishing her toilet she saw 'a violet ball the size of a small orange coming down through the forest', followed by a second, observed them circling round each other, and attempted to investigate, thinking they were some brand new kind of luminous insect – but one went off into the bushes, and the other sank out of her reach in the water – she could see it glowing until it vanished in the depths. The next day, she asked the Fan about the phenomenon; they replied that it was an 'Aku' or devil bush: 'More than ever did I regret not having secured one of those sort of two phenomena. What a joy a real devil, appropriately put up in raw alcohol, would have been to my scientific friends!'

Mary Kingsley exploited to the full her role of eccentric, intrepid naturalist-explorer; and she enjoyed offering a set of challenges to the accepted views. She reminds her reader repeatedly that she is covering the same ground as du Chaillu, with his tall and often discredited stories. Like him, she comes across a band of gorillas; quite unlike him, she does not shoot any of them, but quietly observes their behaviour – until one of her companions sneezes: 'I have seen many wild animals in their native wilds, but never have I seen anything to equal gorillas going through bush; it is a graceful, powerful, superbly perfect hand-trapeze performance.'

Mary Kingsley compares herself favourably to the Europeans of some West Coast districts 'where the inhabitants are used to find the white man incapable of personal exertion, requiring to be carried in a hammock, or wheeled in a go-cart of a Bath-chair about the streets of their coast towns'. She makes it very clear that these effete European males are English, for the context of this passage is the French Congo, where you find a very different set of white men. If a native were to say to one of the French, 'You shall not do such and such a thing', it would mean that it would inevitably be done. She values the French above the English, and the Liverpool or Bordeaux trader above the missionary or the government official. She respects the Fan above other peoples, 'for with all their many faults and failings, they are real men'.

For all her inconsistencies, Kingsley began to think and to write about Africa and Africans from a different perspective to many of her predecessors in West Africa, and certainly in sharp divergence from current imperialist attitudes. The contrast was all the more startling because she looked so orthodox and 'respectable'. Where Alexandrine Tinne had worn Arab costume and adopted Arab customs, Mary Kingsley strode, or sloshed, through the mangrove swamps in full Victorian skirts, with a hat and umbrella, as though about to produce a visiting card from her handbag. It was her mind, and her opinions, which were unusual. She saw the peoples of Africa, not as an inferior race who needed conversion to bring them out of darkness into the light of European civilisation, but as distinctive, and different. She made a great many sweeping generalisations – she was not working in any systematic way; but, if she did not understand the distinctiveness of even the Fan, she made an astonishing attempt. By the end of her year-long travelling, she had become, as much as a naturalist and collector of fishes, an embryonic ethnologist. As she vividly expressed it, the main obstacles to understanding 'an African tribe in its original state' lie in the whole set of difficulties with your own mind: 'Unless you can make it pliant enough to follow the African idea step by step, however much care you may take, you will not bag your game.' To illustrate this point, she recounts a story of a representative of Her Majesty in Africa out for a day's antelope shooting; always, just before he got within shot, the game saw something and bolted: 'happening to look round, he saw the boy behind him was supporting the dignity of the Empire at large, and this representative of it in particular, by steadfastly holding aloft the consular flag. Well, if you go hunting the African idea with the flag of your own religion or opinions floating ostentatiously over you, you will similarly get a very poor bag.'

Mary Kingsley was largely self-taught, and she had the intellectual and emotional toughness of the individual who has been compelled to reconstruct her own life. She consciously challenged the

29. Mary Kingsley, the epitome of respectability.

prejudices and conventions of the late Victorian period, which she characterised as 'humbug'; and she found the ideal form in which to do it, by writing a travel narrative which simultaneously imitated and mocked the form. She chose to go to the least healthy, least regarded area of Africa. She went, not as an extension of empire (Lady MacDonald) nor even as a missionary (Mary Slessor), but in the guise of a trader, because that was something the African people could understand, and not feel threatened by. But she refused to mimic the behaviour of the great white male hunter like du Chaillu, or indeed her own father, who had slaughtered wildlife wholesale on his travels: she deliberately dressed as a woman, and would brandish her umbrella at a hippopotamus or whack a crocodile on the head with a paddle. She describes an incident with a leopard:

> On one occasion a big leopard had attacked a dog, who, with her family, was occupying a broken-down hut next to mine. The dog was a half-bred boarhound, and a savage brute on her own account. I, being roused by the uproar, rushed out into the feeble moonlight thinking she was having one of her habitual turns-up with other dogs, and I saw a whirling mass of animal matter within a yard of me. I fired two mushroom-shaped native stools in rapid succession into the brown of it, and the meeting broke up into a leopard and a dog. The leopard crouched, I think to spring on me. I can see its great, beautiful, lambent eyes still, and I seized an earthen water-cooler and flung it straight at them.
>
> It was a noble shot; it burst on the leopard's head like a shell and the leopard went for bush one time.

This incident is prefaced by Mary Kingsley's tongue-in-cheek statement, 'I am habitually kind to animals, and besides I do not think it is ladylike to go shooting things with a gun', and a closing disclaimer: 'Do not mistake this for a sporting adventure.' She manages both to disparage the traditional close encounter, which would have ended with the leopard's death, and to convey the effectiveness of a no-nonsense woman's approach, the well-directed pottery calabash method. Her technique in the bush, when meeting a potentially dangerous animal, is to melt into the track; on another occasion, she records coming upon a leopard in a violent storm: 'he was crouching on the ground, with his magnificent head thrown back and his eyes shut. His fore-paws were spread out in front of him and he lashed the ground with his tail.' In the face of danger – the tornado, not the human being – that 'depraved creature' swore. Mary Kingsley ducked back behind the rocks, and after what seemed like twelve months, but was more likely twenty minutes, the leopard made off. Instead of a trophy, Mary Kingsley brought back images of the animals in their natural habitats; just as she creates convincingly realistic and certainly unfamiliar snapshots of the West African people.

Seeking a 'respectable' role, for home consumption, she posed as a scientist. Even without formal education, she seemed to imply, this was something anyone could do with a modicum of ingenuity and courage. But as an ethnologist she shows a distinctly unjudgmental and unrespectable curiosity not just in 'fetish' but in cannibalism, infanticide, witchcraft, and the sexual mores of the coast people. Without ever concealing aspects of behaviour which she found alarming, she contrived to write about the Delta people in the same terms she might use to describe English society; for example, when outlining the circumstances in which Mary Slessor rescued a mother who had given birth to twins, she adds, about the length of time of the woman's future ostracism: 'I tried to find out whether there was any set period for this quarantine, and all I could arrive at was that if – and a very considerable if – a man were to marry her and she were subsequently to present to society an acceptable infant, she would be to a certain extent socially rehabilitated, but she would always be a woman with a past – a thing the African, to his credit be it said, has no taste for.' Suddenly, through that European staple of fiction and social drama, the woman with a past, the treatment of twins in the Niger Delta becomes a custom to be understood, though not approved, while at the same time Mary Kingsley surreptitiously introduces another tribal subject, the English upper-middle classes, as a rich field for the ethnologist.

Mary Kingsley climbed Mount Cameroon, 'The Throne of Thunder', on the way home. She was consciously following in the steps of Sir Richard Burton, but she could claim to be the first European to get there by the *south-east* route. She was also, of course, the first woman.

She returned to England as a public figure, in demand for talks and lectures, and the publication of her book in 1897 sharpened her profile. She entered into public debates and controversies about the liquor traffic, and about colonial government, striving to win her arguments by force of logic backed up by first-hand experience. Her most distinctive contribution, though, is the unusual perspective from which she approached Africa and Africans, one which she labelled 'scientific', and one which was remarkably free from bias and prejudice. Without close family, she seems to have enjoyed the society of West Africans quite as fully as any she had experienced in Cambridge or London. In England, she found herself part of a generation whose utterances, at least, were 'tainted with humbug'. In Africa, she discovered a powerful empathy with the landscape: 'It is the non-human world I belong to myself,' she wrote to a close friend. 'My people are mangroves, swamps, rivers and the sea and so on – we understand each other.' She preferred Africa to England, and night to day, as during her nights on the Rembwe, that 'great, black, winding river with a pathway in its midst of frosted silver where the moonlight struck it: on each side the ink-black mangrove walls, and above them the band of star and moonlit heavens that the walls of mangrove allowed one to see'. She survived West Africa, the 'white man's grave'. In 1900 she sailed out to South Africa, looking for 'odd jobs'. She was 'down in the ruck of life again', glad to be away from London society and from politics, that gateway into which she had 'so strangely wandered'. In the Simonstown camp she nursed Boer prisoners of war. 'All this work here,' she wrote, ' the stench, the washing, the enemas, the bed-pans, the blood, is my world.' In spite of characteristic precautions – she smoked to ward off infection, and drank wine instead of water – she died in June 1900, and was, at her last request, buried at sea.

Marianne North's travels form a strange and wonderful coda to the work of the naturalists and plant-hunters, though her detailed knowledge of plants qualified her to rank as a naturalist as well as a painter: four hitherto unknown species were named after her. The fruits of her successive journeys hang in her own monument, the gallery designed to her specification – and built with her own money – in Kew Gardens: a visual record of all the plants she had seen and painted, plants, in her own words, 'in their homes'. The artistic effect is almost overpowering; Wilfrid Blunt described the collection as a 'gigantic botanical postage-stamp album'. The impact on the mind is unforgettable.

This great life's work began almost accidentally. Born in 1830, Miss North was an accomplished Victorian woman, with a lovely singing voice and a facility for painting, who devoted herself to the care of her father after her mother's death in 1855. He took her and her sister to Europe – Switzerland, Austria, Spain, Italy, Greece, even to Egypt and Baalbek, where she sketched and visited the sites. John Addington Symonds saw the sisters at Murren: 'The eldest was blonde, tall, stout, good-humoured, and a little satirical. The second was dark and thin and slight, nervous and full of fire and intellectual acumen.' He married Catherine, the younger, leaving Marianne to accompany her ageing and ailing father. On his death in 1869, she was shattered. 'For nearly forty years he had been my own friend and companion, and now I had to learn to live without him, and fill up my life with other interests as best I might.' As soon as she decently could she went to Menton, where she tried 'to learn from the lovely world which surrounded me there how to make that work henceforth the master of my life'. Her family were concerned about her. Her brother-in-law thought her unhappy: 'There seems to be no future for her and she goes on painting in a ceaseless, mechanical sort of way as if to fill her time up.'

But, as with Mary Kingsley, the death of a parent seemed to release in her springs of energy and independence. She was wealthy, and well-connected: her father, a member of parliament and from a long-established family with roots in Norfolk, had ensured that she had a wide circle of friends. She had stayed with the Francis Galtons, and Sir William Hooker had presented her with a hanging bunch of *Amherstia nobilis*. With her father she had enjoyed a privileged, comfortable but far from sheltered life, equally at ease in the cities of Europe or on strenuous walks at high altitude in the Alps. In England she became restless, impatient with the eternal dressing up and stuffiness, bored with the Norfolk conversation about turnips and partridges. She longed to travel further afield; Charles Kingsley's account of his travels in Brazil and the West Indies in 1871, *At Last*, 'added fuel to the burning' of her rage to see the tropics. (Kingsley supplied letters of introduction for her to Brazil and the West Indies – Marianne North was a brilliant networker, and always travelled armed with influential contacts.) A chance invitation led to a trip to the United States in 1871, where she met the young Mrs Agassiz, the wife of the much older Swiss naturalist. Mrs Agassiz, who had been up the Amazon with her husband, as she described in *Journey to Brazil* in 1868, showed her all sorts of treasures at the university at Cambridge, Massachusetts – palms, butterflies and corals – and agreed with her that 'the greatest pleasure we knew was to see new and wonderful countries'. Before long, Marianne North was heading for the warmth of Jamaica. She found a house for rent, half-hidden amongst the foliage of the long-deserted botanical gardens of the first settlers, and hired it for £4 a month.

> From my verandah or sitting-room I could see up and down the steep valley covered with trees and woods; higher up were meadows, and Newcastle 4000 feet above me, my own height being under a thousand above the sea. The richest foliage closed quite up to the little terrace on which the house stood; bananas, rose-apples (with their white tassel flowers and pretty pink young shoots and leaves), the gigantic bread-fruit, trumpet-trees (with great white-lined leaves), star-apples (with brown and gold plush lining to their shiny leaves), the mahogany-trees (with their pretty terminal cones), mangoes, custard apples, and endless others, besides a few dates and cocoa-nuts. A tangle of all

> sorts of gay things underneath, golden-flowered allemandas, bignonias, and ipomoeas over everything, heliotropes, lemon-verbenas, and geraniums from the long-neglected garden running wild like weeds: over all a giant cotton-tree quite 200 feet high was within sight, standing up like a ghost in its winter nakedness against the forest of evergreen trees, only coloured by the quantities of orchids, wild pines, and other parasites which had lodged themselves in its soft bark and branches.

She painted all day, going out at dawn and retreating indoors at noon, when she 'worked' at flowers during the afternoon rain-storms; then, when the skies cleared, she would walk up the hill to explore a new part of her paradise, returning home in the dark. Society found her, even in her garden retreat, but she soon learned to avoid the balls and 'heavy' parties of colonial life. She went back to London for a short break, and then sailed for South America, the focus of her original impulse, to continue her studies of tropical plants. For a while she lived in Rio, but was happier when she ignored well-meant advice and warnings and made her way to Morro Velho:

> Just below the flower-garden was a perfect temple of bananas, roofed with their spreading cool green leaves, which formed an exquisite picture. Sometimes a ray of sunlight would slant in through some chink, and illuminate one of the red-purple banana flowers hanging down from its slender stem, making it look like an enchanted lamp of red flame. Masses of the large wild white ginger flowers were on the bank beyond this temple, and scented the whole air. Farther down the steep path were masses of sensitive plants covering the bank with the brightest of green velvet and delicate lilac buttons.

For her, the gardens of the tropics, and the surrounding forest into which they imperceptibly merged, were temples; and the sense of wonder, worship almost, which they evoked in her comes almost as strongly from her writing as from her painting.

Marianne North's independence was established on this first journey to the interior. She travelled relatively rough – she seems not to have minded discomfort, so long as she was warm – and rather relished the fact that, back in the city of Petropolis, it was thought 'rather shocking and dangerous' for her to wander over the hills alone.

After each trip she would return to London; but she now found the winters unbearably cold, and before long she was planning a fresh expedition – a relatively brief visit to Tenerife in 1875, where she dutifully took in the famous view of the Peak described by

30. Marianne North at work.

Humboldt. But then she was really launched: asked by some fellow guests at a country house party about her next destination, she replied, vaguely, Japan, and was invited politely to join them. A fortnight or so later she was heading west across the Atlantic. She took the train to Chicago and Salt Lake City, and then the stagecoach, and was soon installed in the Mariposa Grove with her easel, painting the great trees. She moved on from Yosemite to Japan, where she obtained a special order from the Mikado to sketch for three months in Kyoto. She loved Japan, but the intense cold undid her: 'After sketching all day amongst the dead leaves, and morning white frosts, I used to be scarcely able to stand from stiffness and coming rheumatism, and had to hold on by a tree at first, till I could use my feet.' She journeyed on, sooner than she had planned, via Hong Kong and Saigon to Singapore, Borneo, Java and Ceylon.

Perhaps it was in Borneo, which she reached early in 1876, that Marianne North came closest to the heart of the natural world. Borneo was a home to the orchid, the pitcher plant and the orang-utan, three of the species which most entranced the West. Like Wallace some twenty years before, she was a guest of the Rajah (Sir Charles Brooke, who had succeeded his uncle) at Kuching, and had the benefit of his protection and staff; she became a good friend of the Ranee, who in her memoirs gives a portrait of her: 'tall, lean [in contradiction to Symonds] and fair; her nose was rather large, her lips were rather thin, she wore blue spectacles and was not good-looking'. The younger woman was clearly taken aback by this friendly but commanding visitor, who had come equipped with one small portmanteau, but case after case of finished pictures, canvases and painting materials, and three easels, and who refused the invitation to a siesta after *déjeuner* – 'Rest? How very odd!' – and immediately started turning the second spare room into a studio. Then, after afternoon tea at half-past four, she accompanied the Ranee into a canoe, well equipped for the jungle in short petticoats, light woollen jacket and Wellington boots, and they set off in search of some pitcher plants, which were cut down to her instructions and taken back to Kuching, ready for work the next morning. Then, as dusk fell promptly at six, they returned for dinner, after which the Ranee played a Beethoven sonata – *all* of it, she was ordered – and after that Miss North took her place at the piano and sang, and sang; finally the Ranee, three of whose children had died on a homeward voyage from drinking a tin of poisoned milk, but who still had 'one small tyrant of eighteen months', was allowed to go to bed. It was a relief to her to find that her guest did not need much entertaining, so long as she was supplied with enough plants to draw. In spite of being patronised and called 'child', she warmed to Marianne North, although the Rajah found himself corrected rather too sharply for his liking on the proper care of orchids.

The arrival of an official in need of a bed prompted her suggestion that she should go to the Brookes' mountain-farm for a while.

> The forest was a perfect world of wonders. The lycopodiums were in great beauty there, particularly those tinted with metallic blue or copper colour; and there were great metallic arums with leaves two feet long, graceful trees over the streams with scarlet bark all hanging in tatters, and such huge black apes! One of these watched and followed us a long while, seeming to be as curious about us as we were about him. When we stopped he stopped, staring with all his might at us from behind some branch or tree-trunk; but I had the best of that game, for I possessed an opera-glass and he didn't, so could not probably realise the whole of our white ugliness.

Marianne North was one of the first naturalists to record, rather than collect; most of her predecessors would have been armed with a gun. She met a young man who had been sent out as a naturalist by Sir Charles Lyell in search, according to her, of the 'missing link', or men with tails, and was now earning a living by collecting. He brought her some grand specimens of the largest of all pitcher plants for her to paint. The picture prompted Veitch to send out his own traveller for seeds, and Joseph Hooker named the species after her, *Nepenthes northiana.*

By this stage, Marianne North's reputation meant that she travelled increasingly as part of the official ruling caste: judges, governors, even governor-generals and their wives were her hosts and hostesses, and she moved, when it was convenient, with a large retinue of borrowed servants. But she never became reconciled to the bland chatter of the croquet- and tennis-playing classes, and she retained her sense of proportion. In Kandy she slept in the huge state-rooms of the Pavilion, last occupied by the Prince of Wales. She felt oppressed by the dreary grandeur, 'like a sparrow who had by a mistake got into an eagle's nest', and by six the next morning was outside at her easel. In Kalutara, she stayed with Mrs Julia Cameron, and gave an interesting account of her 'constructed' photographic methods:

> She dressed me up in flowing draperies of cashmere wool, let down my hair, and made me stand with spiky cocoa-nut branches running into my head, the noonday sun's rays dodging

31. Marianne North photographed by Julia Margaret Cameron.

> my eyes between the leaves as the slight breeze moved them, and told me to look perfectly natural (with a thermometer standing at 96)! Then she tried me with a background of bread-fruit leaves and fruit, nailed flat against a window shutter, and told *them* to look natural, but both failed.

This is the 'natural' painter, the painter from life, speaking.

After this trip, the officials at Kensington Museum asked for some paintings to exhibit in their galleries; and following her two-year visit to India, Marianne North hired a gallery in Conduit Street for a more extensive exhibition in July and August 1879. The reviewer of the *Pall Mall Gazette* suggested that the paintings should be located at Kew, and it was this comment which prompted her to offer the collection to Sir Joseph Hooker, with a gallery to house them: she wrote to him, characteristically in transit, from Shrewsbury station, where she had missed a train. Charles Darwin advised that she should not attempt 'any representation of the vegetation of the world' until she had seen and painted the Australian. This was like a royal command, and in 1880 she was heading east once more. She painted in Australia, New Zealand (too cold for her) and, her last continent, South Africa. She was beginning to flag, and her health became fragile: she had to turn down a tempting extension of her tour to Madagascar. But she summoned up her resources for a final major expedition, this time to Chile, to paint the one great tree which had so far escaped her in the wild, the *Araucaria imbricata* – the monkey-puzzle tree. But it is her search for the blue puya which seems, in her description, to summarise the energy and intense longing of her years of plant-hunting and painting:

> We tied up the horses when it became too steep, and proceeded on foot right into the clouds; they were so thick that at one time I could not see a yard before me, but I would not give up, and was rewarded at last by the mists clearing, and behold, just over my head, a great group of the noble flowers, standing out like ghosts at first, then gradually coming out with their full beauty of colour and form in every stage of growth; while beyond them glittered a snow-peak far away, and I reached a new world of wonders, with blue sky overhead, and a mass of clouds like sheets of cotton-wool below me, hiding the valley I had left.

The North Gallery was opened by Hooker on 9 July 1882. The lower walls were lined with 246 different kinds of wood which Marianne North had collected, a shrine to the world's trees. The 832 paintings on her travels, largely but not exclusively of plants, are a tribute to the rich flora she had seen and recorded where they

32. Marianne North's painting of the blue puya in the Chilean Andes. (Marianne North Gallery, Royal Botanic Gardens, Kew)

grew: in botanic gardens and collections, but also in the jungle and on mountain ridges. As Hooker commented, some of the plants she had painted were already disappearing.

Marianne North's *Recollections* were sent, following Hooker's advice, to John Murray, who commented, reasonably, on their length and, more curiously, on 'their very peculiar character'. Macmillan's also felt they required editing, but Marianne North was not well enough to undertake the task, which fell, after her death in August 1890, to her sister. They were published in 1892. A century later, their half-indiscretions and discursiveness, and their aura of uncompromising honesty, still convey an unusual freshness and distinction. Marianne North's status puzzled her contemporaries: her sister commented that she was 'no botanist in the technical sense of the term: her feeling for plants in their beautiful living personality was more like that which we all have for human friends. She could never bear to see flowers uselessly gathered – their harmless lives destroyed.' W. Botting Hemsley, the assistant for India at Kew, had checked the botanical details of the book for errors; and, in a memorial essay, he called her, 'not a botanist', but 'a great observer'. Marianne North had acquired a scientific frame of mind, and was at least an agnostic; but she conveys in her writing, and in her painting, a sense of awe in response to the richness and diversity of the plants she recorded. She was a non-intrusive traveller and a non-collecting collector, providing an alternative approach to the ruthless, large-scale despoliation of the mid-century plant-hunters. Turning nature into art, she then enshrined her vision of the natural riches of the world in a special temple on the sacred botanical site of Kew.

9 *A New Mythology*

THE SCIENTIFIC TRAVELLERS CHANGED the British, and European, view of the world. The hard core of their collective discoveries was represented by the theory of natural selection, with the figure of Darwin as prophet, and *The Origin of Species* as the sacred book. In the popular imagination, the central message which could be extracted was that man was descended from the animals, and the idea of man as primitive, savage, became dominant. If all men were savage, then it was necessary for the more civilised to control the rest, or, put another way, it was natural for the strongest and most cunning to rule. By the end of the century, there was a radically altered perspective on the origins of humanity, which swivelled to look, even more disturbingly, towards humanity's future. Regardless of whether or not people reconciled the new perspective with belief in God – and some responded by denouncing it as a heresy – an unsettling set of questions was opened up. If God was not in charge, or even if God was only in charge through an infinitely remote and unpredictable mechanism, did human life fall within any understandable form of control? What did the theory of the survival of the fittest mean in terms of relationships – with other classes, or with other races? What were the links with the other orders of apes, and with the rest of the natural world? Could man, with the gift of consciousness and the ability to predict, intervene and actively shape a better future?

The past, and the distant past, were studied as never before. Man's evolution, and each step in its mysterious and bewildering path, came under intense and often highly speculative scrutiny, a strange mixture of scientific investigation and myth-making. Inevitably, the institutions and customs which seemed to validate and justify the existing structures of society were themselves anatomised: if the order which underpinned the Church and the established political and social systems of, say, Victorian Britain was suspect, then those institutions themselves might need to change. The dead hand of conformity loosened its grip. Shakespeare's Edmund could cry, 'Thou, Nature, art my goddess': 'Now, gods, stand up for bastards!', and not be challenged.

As the century drew on and approached the last decade, with all the extra spin imparted by notions of a *fin de siècle*, the urgency of the questions, the force with which they were put, increased. The questions surfaced, inevitably, in all kinds of cultural and literary forms, in the novel and short story, and in drama. Drama, because of its peculiarly public form, and the active intervention of the censor, lagged slightly behind. Ibsen's plays of the 1880s, with their devastating assaults on convention, and their unremitting exploration of ideas such as inheritance, won only a minority hearing on the London stage. *Ghosts*, with its references to syphilis, was labelled obscene. The two British playwrights with the intellectual firepower to enter the arena were, significantly, both Irish. Wilde, who did address the questions directly in *The Soul of Man under Socialism*, opted to undermine the tribal conventions of the English establishment more obliquely in his society comedies, though the concepts of the modern, the new morality and the ideal permeate his dialogue. The altered perspectives came like manna to feed Shaw's voracious appetite for ideas; but it took him some time to establish an audience for his post-Darwinian plays, and only in the first decades of the twentieth century did he tackle the issues head on, in, for example, *Man and Superman*. Ideas of empire were central to the public spectacles of Victorian theatre, but the awkward issues it raised tended to remain just off-stage in late Victorian drama. Jack Worthing plans to send his wicked younger brother Ernest to Australia. The colonies were places from which characters might make a sudden convenient return, like Captain Ardale in *The Second Mrs Tanqueray*, marvellously transformed by some heroic act, or to which they could be dispatched, as in Henry Arthur Jones's *The Liars*. The last scene is the drawing-room in the Victoria Street flat of Colonel Sir Christopher Deering (Daring? Derring-do?) – 'The room is in great confusion, with portmanteau open, clothes, etc., scattered over the floor; articles which an officer going to Central Africa might want are lying about.'

Charles Kingsley's *The Water Babies*, which began serialization in the August 1862 number of *Macmillan's Magazine*, was one of the first literary explorations of natural selection. Kingsley's Christian socialism, and his bias towards evolution, as well as his own experience as a naturalist, made him an enthusiastic follower of Darwin. He was welcomed into the fold by Huxley, who thought him a 'very real, manly, right minded parson', adding: 'but I am inclined to think on the whole that it is more my intention to convert him than his to convert me'. Kingsley wrote straight to Darwin after reading *The Origin of Species*: Darwin was right, he must give up much that he had believed; and he was prepared to embrace a conception of Deity who 'created primal forms capable of self-development'. Darwin accepted Kingsley's formula gratefully, and incorporated it into the last chapter of the second edition. Like Asa Gray, Kingsley believed that the tendency of physical science was 'not towards the omnipotence of Matter, but to the omnipotence of Spirit'; and when Bates discovered the Mocking butterflies in the Amazon, Kingsley wrote to congratulate him. Bates's explanation, though the simplest, was the most wonderful of all, because it looked most like an immensely long chapter of accidents, each testifying to divine providence.

The Water Babies, one of those Victorian children's books which work differently but perhaps even more effectively for adults, is a coded tour round the scientific debates of the mid-century. 'I have tried, in all sorts of queer ways,' Kingsley explained to F.D. Maurice in 1862, 'to make children and grown folks understand that there is a quite miraculous and divine element underlying all physical nature. . . . And if I have wrapped up my parable in seeming Tomfooleries, it is because so only could I get the pill swallowed by a generation who are not believing with anything like their whole heart, in the Living God. Meanwhile, remember that the physical science in the book is *not* nonsense, but accurate earnest, as far as I dare speak yet.' Tom, the chimney-sweep's apprentice, traces his evolutionary path backwards, as he slips into the cool clear water of the Pennine limestone stream in Vendale; naturally, to live in the element of water he acquires a set of external gills round the 'parotid region of his fauces'. Thus equipped, he is in an excellent position to explore the natural world via the water, and from an eft progresses both physically and morally (the two go together in Kingsley's complex parable) all the way to full-blown Victorian manhood, at which point he is of course rewarded by the company of Miss Ellie, the golden-haired little girl whom he first scared when he tumbled out of the chimney in Harthover House. Tom has evolved into a great man of science and industry, who can plan railroads, and steam engines, and electric telegraphs, and, in keeping with the age, rifled guns.

The idea of man's evolution, and especially of the critical stage from ape to man, is one of the story's main threads. It begins as a metaphor. When Tom first sees Ellie, who looks to him like an angel out of heaven, he suddenly catches sight of his image in a mirror: 'What did such a little black ape want in that sweet young lady's room?'; Ellie screams as loud as a peacock, and the hunt is up: 'And all the while poor Tom paddled up the park with his little bare feet, like a small black gorilla fleeing to the forest. Alas

33. Linley Sambourne's illustration of Owen and Huxley inspecting a water-baby. (*The Water Babies*, 1886 edition)

for him! there was no big father gorilla therein to take his part; to scratch out the gardener's inside with one paw, toss the dairymaid into a tree with another, and wrench off Sir John's head with a third, while he cracked the keeper's skull with his teeth, as easily as if it had been a cocoa-nut or a paving-stone.' Tom, through society's fault, is more animal than human. (His father is dead, and his mother is in Botany Bay, and he is the slave of the degenerate, industrial, alcoholic, violent Grimes.) As mother nature, alias the queen of the fairies, Mother Carey, the Irishwoman, Mrs Bedonebyasyoudid/Mrs Doasyouwouldbedoneby, says to the fairies of the stream: 'He is but a savage now, and like the beasts which perish; and from the beasts which perish he must learn.'

Tom has a number of narrow escapes on his journey – twice, he is almost gobbled up by an otter. More alarmingly, when chatting with a lobster in a rock-pool, he gets caught in Professor Ptthmllnsprts' net, and only avoids being given two long names and made into a speech for the British Association by biting the professor's finger very hard. (This composite scientist – Huxley? Owen? – can be mocked by Kingsley because he is passed off as of foreign descent, though with vowels added he can be identified as the ubiquitous Put-them-all-in-spirits.) Several topical scientific controversies muscle in on the story, particularly Huxley and Owen's debate over the 'hippopotamus major' test. But Kingsley's deep-rooted trust in science prevails. On Tom's journey backwards, he sets off for Mother Carey's pool, which Kingsley places in the Arctic Circle, first in the company of the petrels, then with the mollys. (On the way, they pass the wrecks of many a gallant ship, some 'with the seamen frozen fast on board', all true English hearts, searching for the white gate of the North-West Passage – though, as the mollys tell Tom, there is never a crack of one, and just as well, or they would have killed every whale in the sea.) Mother Carey is a white marble lady, sitting on a white marble throne. 'And from the foot of the throne there swum away, out and out into the sea, millions of new-born creatures, of more shapes and colours than man ever dreamed.' Tom's version is that she is making new beasts out of old; she corrects him: 'I sit here and make them make themselves.' Mother Carey's advice to Tom is to go backward, and she tells him a version of the Prometheus/Epimetheus story to make the point: the children of Epimetheus follow him in looking *behind* them to see what had happened, till they really learn to know now and then what will happen next; while the children of Prometheus are fanatics and

34. Tom coming down the chimney. (Linley Sambourne)

theorists. The children of Epimetheus, of course, are hard-working, practical scientists.

Kingsley even includes a composite portrait of a scientific naturalist, whom Tom meets in the land of Hearsay: 'a poor, lean, seedy, hard-worked old giant, as ought to have been cockered up, and had a good dinner given him, and a good wife found him, and been set to play with little children'. 'He was made up principally of fish bones and parchment, put together with wire and Canada balsam; and smelt strongly of spirits, though he never drank anything but water: but spirits he used somehow, there was no denying. He had a great pair of spectacles on his nose, and a butter-fly net in one hand, and a geological hammer in the other; and was hung all over with pockets, full of collecting boxes, bottles, microscopes, telescopes, barometers, ordnance maps, scalpels, forceps, photographic apparatus, and all other tackle for finding out everything about everything, and a little more too.' And, just as important, he was running backwards, as fast as he could. He, too, whips out a bottle and cork, to collect Tom; but Tom dodges between his legs; the giant makes a truce, for, as Kingsley says, 'he was the simplest, pleasantest, honestest, kindliest old Dominie Sampson of a giant that ever turned the world upside down without intending it'. If the giant physically brings Wallace to mind, the description of turning the world upside down sounds very like Darwin, as does the giant's motto: 'Do the duty which lies nearest you, and catch the first beetle you come across.'

But Kingsley was a clergyman, and he was bound to integrate the knowledge of science, drawn from beetles and bats and efts, into a moral and spiritual system, rather than rely on doctrines of *laisser-faire* or brute competition. Tom is given a chilling insight into what happens to people who only do what is pleasant, by means of a wonderful waterproof book full of colour photographs, called 'The History of the great and famous nation of the Doasyoulikes, who came away from the country of Hardwork, because they wanted to play on the Jews'-harp all day long'. These people regress, back first to the root- and nut-gathering days; they start to live in trees, to stay out of the lions' way, they grow uglier, and more hulking – the ladies will only marry the strongest and fiercest gentlemen – their feet change shape, they become hairier; they become fewer, and finally Tom sees that they cannot walk upright: '"Why," he cried. 'I declare they are all apes."' And in the next 500 years they were all dead and gone, by bad food and wild beasts and hunters, except one tremendous old fellow, seven feet high. Kingsley gave him a dreadful fate, at the hands of a rather ungentle pseudo-scientific explorer:

> M. du Chaillu came up to him, and shot him, as he stood roaring and thumping his breast. And he remembered that his ancestors had once been men, and tried to say, 'Am I not a man and a brother?' but had forgotten how to use his tongue; and then he had tried to call for a doctor, but he had forgotten the word for one. So all he said was, 'Ubboboo!' and died.

If only they had behaved like men, and set to work to do what they did not like, they could have been saved from being apes; for nature, says Kingsley in the person of the 'fairy', can make beasts into men, by circumstance, and selection, and competition; but: 'if I can turn beasts into men, I can, by the same laws of circumstance, and selection, and competition, turn men into beasts'. Fortunately, however, Tom had made up his mind to go on a journey and see the world, like a true Englishman. Kingsley tackles the moral dimension head on: he offers a simple solution to a complicated set of problems, a mixture of hard work and doing what you don't particularly want to – in Tom's case, going to the help of Grimes. But the question of how to behave in an apparently random world, 'the absolute empire of accident', is not just a dilemma for an orthodox, if enlightened clergyman. It is the same crux which Darwin's bulldog, Huxley, confronted. In *Evolution and Ethics*, he argued that 'The ethical progress of society depends not on imitating the cosmic process . . . but in combating it.' This was a difficult position for an orthodox Darwinian to adopt, but Huxley challenged the view that fitness could be equated with goodness.

Another early literary response to Darwin came from Samuel Butler in 1872, with *Erewhon*, like *The Water Babies* a journey both backwards in time and into the future, in search of a new utopia. Butler was a kind of embryonic appendage to Darwin. He was educated, like Darwin, at Shrewsbury and Cambridge; his father had even hunted beetles with Darwin in the Welsh mountains. He, too, went through a rite of passage in a long voyage to the uttermost ends of the world, though in his case he sailed to New Zealand to escape from his family. In the remoteness of the Southern Alps, he read his mentor's great work, and it became for him, for a time, his book of books. Out of his reactions came a number of articles for the Christchurch Press, including 'Darwin among the Machines', which on his return to England he used as part of his utopian satire. *Erewhon* confronts a great many myths and monsters, including Victorian religion and education – Butler reserved his assault on that other unnatural institution, the

'Victorian family', for *The Way of All Flesh* – but one of the most striking elements is the journey of exploration through perilous territory to a new country. Butler cannot have known that New Zealand was one of the last inhabitable islands in the world to be colonised by man *per se*, not just by 'white man', so that to set his imaginary state on the far side of the Southern Alps, as far away from industrial Europe as it is possible to get, was an inspired guess. In *Erewhon*, the march of progress has been halted, the clocks put back: predicting that man may be driven out by machines and lose the struggle for existence, the Erewhonians have taken matters into their own hands and dismantled all but the simplest.

Butler can be seen as part of the great outward bound movement of the mid-century, in which Britain, like most of Western Europe, spilled out to investigate, explore, colonise and exploit the rest of the world. Butler, though perhaps a special case, was a kind of literary shadow to the scientific naturalists, as well as a compendium of all the other aspects of the British diaspora. He was a semi-scientific traveller and explorer, discovering, while he hunted for good sheep country, the pass from the Canterbury plains across the Southern Alps, which he made the basis for his protagonists' journey into Erewhon; he was a settler, who did not stay; he came back rich from raising sheep – though, being Butler, he lost all his money. He was also an anti-missionary, preaching Darwin at this stage of his paradoxical life instead of the Anglicanism of his birthright. Being the son of a parson-naturalist, he sent back dried ferns for his father's herbarium. Much to Butler's later annoyance, Darwin shaped him, and his career, in a very direct way. First Butler sought Darwin's approval, sending him his books and even being invited to stay at Down. Then he turned against him, as he had turned against his father, and spent a decade writing against him, with a bitterness that obscured the shafts of insight which illuminated his books on evolution, insights appreciated by Alfred Wallace, and, later, Bernard Shaw.

Butler was, for all his radical questioning of inherited values and systems, a capitalist. Men prospered by luck, or cunning; and the best luck would be to be born warmly wrapped up in high-denomination Bank of England notes. The capitalist base of British imperialism lies at the heart of English society, and of the fiction which reflects that society, as Edward Said has shown in *Culture and Imperialism*. Behind the orderly parkland and shrubbery walks of Mansfield Park lie the plantations of Antigua; at Rochester's shoulder stands the spectre of his 'black' mad wife. Said quotes the opening of *Dombey and Son*:

> The earth was made for Dombey and Son to trade in, and the sun and moon were made to give them light. Rivers and seas were formed to float their ships; rainbows gave them promise of fair weather; winds blew for or against their enterprises; stars and planets circled in their orbits, to preserve inviolate a system of which they were the centre.

The nineteenth-century world was made for the English merchant; and poor Mrs Dombey, having produced a son and heir to the business, promptly expired. The succession, and the commercial succession – for daughter Fanny, being a girl, did not count – had been achieved; and Mrs Dombey, her genetic task done, was dispensable. The business machine, the house of Dombey and Son, was recharged. Dickens sets it firmly in the centre of the City of London, close to the Royal Exchange, the Bank of England with its vaults of gold and silver, and round the corner from East India House, 'teeming with suggestions of precious stuffs and stones, tigers, elephants, howdahs, hookahs, umbrellas, palm trees, palanquins, and gorgeous princes of a brown complexion sitting on carpets, with their slippers very much turned up at the toes. Anywhere in the immediate vicinity there might be seen pictures of ships speeding away full sail to all parts of the world; outfitting warehouses ready to pack off anybody anywhere, fully equipped in half an hour; and little timber midshipmen in obsolete naval uniforms, eternally employed outside the shop doors of nautical instrument-makers in taking observations of the hackney coaches.' Dickens, alongside the undercutting irony, gives a vivid picture of the energy of commerce, and incidentally sketches in a whole set of businesses which supply the travellers and merchants alike. London's riches depended on the precious stuffs and stones of the East; and a kind of glamour, borrowed, or stolen, from the world across the sea, attached itself to everyone associated with the great web of business and transactions which had its centre in the heart of the City.

The rest of the world, and especially the Empire and the colonies, looms like a shadowy backdrop to the nineteenth-century British novel. Dickens's writing, so alive to the patterns of commerce, offers many examples: *Great Expectations*, *Hard Times*, *Bleak House* with its caustic reference to Mrs Jellyby's obsession with the Niger project. Going abroad was such a common social phenomenon that novelists and dramatists put it to all sorts of uses: a convenient device for getting rid of characters, or allowing them to come back unexpectedly, rich, ruined, mad, or perhaps heroic and amazingly transformed after a thorough moral shaking-up. These characters were given occupations in business, or government, as soldiers or, more rarely, missionaries. But the rather different concept of a serious scientific purpose begins to appear in the early nineteenth century.

One of the earliest striking examples comes from an Irish context, in Maria Edgeworth's novel of 1809, *Ennui*. Cecil Devereux and his wife-to-be Lady Geraldine are presented as the most intellectually and morally independent characters in the story. Devereux spends his days shut up in his own apartment, immersed in the study of the Persian language: 'When he was not studying, he was botanizing or *mineralogizing* with O'Toole's chaplain': all in preparation for a hoped-for appointment in India. Devereux, a philosopher, offers advice and encouragement to his friend Glenthorn:

> It is difficult to judge what are the natural powers of the mind, they appear so different in different circumstances. You can no more judge of a mind in ignorance than of a plant in darkness. A philosophical friend told me, that he once thought he had discovered a new and strange plant growing in a mine. It was common sage; but so degenerated and altered, that he could not know it: he planted it in the open air and in the light, and gradually it resumed its natural appearance and character.

The idea of the open air, and the light, the impetus towards experiment, the warning against premature judgement, the importance of knowledge, all point towards the great burst of scientific exploration which was to follow. Devereux and Lady Geraldine duly have their wish granted, and from India exercise a benign long-distance influence on Glenthorn's search for identity.

A more extensive portrayal of the scientific traveller comes in Elizabeth Gaskell's *Wives and Daughters*, which began to appear in 1864, but which is set, broadly, in the late 1820s and early 1830s. Mrs Gaskell was related to Darwin, and had many friends in the scientific circles of Manchester and Edinburgh. Steady, slow Roger Hamley, the squire's younger son, is gradually revealed as a developing and far from obvious ideal: as a boy, he is fond of natural history, which takes him out of doors, and when he is in the house he is 'always reading scientific books that bear upon his pursuits'. To everyone's amazement, it is Roger, not his fastidious poetry-writing brother Osborne, who becomes Senior Wrangler in the Cambridge Mathematical Tripos, and goes on to publish a scientific paper in answer to a great French physiologist; who is invited on the strength of that to meet the zoologist Geoffroi

Saint-Hilaire; who is chosen to make a scientific voyage 'with a view to bringing back specimens of the fauna of distant lands, and so forming the nucleus of a museum'. The journey Hamley makes is clearly a parallel to that of Darwin, with the continent of Africa replacing South America. An early outline describes him: 'Roger is rough, & unpolished – but works out for himself a certain name in Natural Science, – is tempted by a large offer to go round the world (like Charles Darwin) as a naturalist.' Hamley moves south from Egypt, past Abyssinia towards the Cape, and is kept before the reader's attention by a series of intermittent references: a letter telling of a bout of fever, a report in a scientific journal of the annual gathering of the Geographical Society: 'Lord Hollingford had read a letter he had received from Roger Hamley, dated from Arracuoba, a district in Africa, hitherto unvisited by any intelligent European traveller'.

Mrs Gaskell died with the novel not quite completed. Frederick Greenwood, her editor at the *Cornhill*, sketched in the end of the story: Roger Hamley returns from Africa to marry Molly Gibson, becomes a professor at a great scientific institution, 'and wins his way in the world handsomely', like Huxley. Significantly, Greenwood uses a botanical simile, and a biological explanation, to account for the likeness in unlikeness between the two brothers. Mrs Gaskell, he argued, had the people in her story 'born in the usual way, and not built up like the Frankenstein monster; and thus when Squire Hamley took a wife, it was then provided that his two boys should be as naturally one and diverse as the fruit and the bloom on the bramble. "It goes without speaking." These differences are precisely what might have been expected from the union of Squire Hamley with the town-bred, refined, delicate-minded woman whom he married.' Mrs Gaskell, writing in the aftermath of *The Origin of Species*, provided in the personality and the career of Roger Hamley a model for the new society: a man whose temperament and upbringing made him suited to science, a science founded on the knowledge of plants and clarified by systematic exploration; she even promoted science as a legitimate and properly rewarded profession.

By the end of the century, novelists such as Conrad, Kipling, Stevenson and Rider Haggard were placing the 'savage' world squarely in the foreground. Conrad was the most penetrating and disturbing transcriber of the imperial and commercial process. After all, he had lived it at first hand. Working for British companies, he had, in addition to the gifts of an artist, the distance and objectivity which comes from being an outsider. He had served, as midshipman to mate and finally master, on the ships which made Dombey and Son's wealth, battering his way across the Indian Ocean and round the coasts of Malaysia, Sumatra, Borneo, New Guinea and Australia. It was ships like those Conrad sailed on – the *Highland Forest*, the *Vidar*, the *Otago* – which deposited Wallace at some small port, and then steamed off over the horizon while Wallace made his way inland with his collecting gear. Conrad, like Wallace, had time to reflect on the extraordinary interaction of the white traders with the inhabitants, of the colonial powers with the native rulers and societies. Again and again, his stories and novels lay bare the deep distrust of the process he had witnessed. In his 1898 collection, appropriately entitled *Tales of Unrest*, is the story 'Karain'. One of the earliest appraisals of Conrad's work was offered by Hugh Clifford, colonial administrator and himself an author – of *Studies in Brown Humanity*, among many other books. Clifford was sympathetic, but critical of Conrad's grasp of detail and his understanding of an alien culture. 'Karain', he commented, could only be called Malay 'in Mr Conrad's sense'. Conrad accepted some of Clifford's criticisms, though maintaining that most of the actions and details had come from stories he had personally heard, or which he had checked in books such as Wallace's *Malay Archipelago*.

'Karain: A Memory' is a classic confrontation between the 'civilised' and the 'savage', and is narrated by an Englishman who encountered Karain, the Bugi chief, in his own earlier career as a white gun-runner. Like many of Conrad's stories, it has a complicated framework, in which the reader is taken further and further back to the core story, a tale of Karain's inner life, his mission of revenge which turns into a betrayal, and his subsequent haunting by the ghost of the man he betrayed for love. The events are fantastic, far-fetched, made more powerful for being told by the fugitive Karain to three cynical Europeans at night on a schooner anchored in an island bay. But the details, and the ironic perspective, are perhaps more telling than the story itself, which represents the element of mystery, of the unknown, so missing from the consciousness of the rational Europeans. The incident which triggered Karain's mission of revenge was twofold: the seizure of power from the island chiefs by the Dutch, and the subsequent taking of a Bugi girl by a Dutch trader.

> One day I saw a Dutch trader go up the river. He went up with three boats, and no toll was demanded from him, because the smoke of Dutch warships stood out from the open sea, and we were too weak to forget treaties. He went up under the promise of safety, and my brother gave him protection. He said he came

to trade. . . . His face was red, his hair like flame, and his eyes pale, like a river mist; he moved heavily, and spoke with a deep voice; he laughed aloud like a fool, and knew no courtesy in his speech. He was a big, scornful man, who looked into women's faces and put his hand on the shoulders of free men as though he had been a noble-born chief. We bore with him. Time passed. Then Pata Matara's sister fled from the campong and went to live in the Dutchman's house. . . .

This incident is like the rape of Helen, and the reversal of Caliban's attack on Miranda. The islands have been taken, the culture disrupted, the way of life will never be the same again, as the schooners slip in and out of the lagoons, with cargoes of rifles and powder, and Karain waits to receive his white visitors on what Conrad describes as a stage: 'the purple semicircle of hills, the slim trees leaning over houses, the yellow sands, the streaming green of ravines' – a 'gorgeous spectacle'; as for Karain himself, 'He summed up his race, his country, the elemental force of ardent life, of tropical nature. He had its luxuriant strength, its fascination; and, like it, he carried the seed of peril within.' Karain is, in Conrad's treatment, nature itself: the savage, the unknown, the unpredictable, infinitely attractive, yet dangerous.

As Karain unfolds his history, he turns to his white listeners for help, to save him from the demon which pursues him. He asks for leave to sail away with them; or at least for a weapon, or a charm, to protect him. Hollis, moved by the story, rummages in his box and manufactures a charm from some of the personal treasures there: a narrow white glove, a dark blue silk ribbon – 'Amulets of white men! Charms and talismans!' – and a Jubilee sixpence with a hole punched near the rim, presented to Karain with the solemn declaration: 'This is the image of the Great Queen, and the most powerful thing the white men know.' Karain accepts the charm; and at dawn returns to the shore, from which the ghost has departed for ever. Then Conrad adds a last framing sequence. Two of the three traders meet accidentally in London, in the Strand, outside a gunsmith's. Jackson wonders aloud – '"I mean, whether the thing was so, you know . . . whether it really happened to him. . . . What do you think?"'

'"My dear chap,"' cries the narrator, '"you have been too long away from home. What a question to ask! Only look at all this."'

Conrad's description of commercial London, the epitome of nineteenth-century urban civilisation, is uncompromising and chilling, a stark contrast to the 'painted scenery' of the archipelago island; a cityscape which is a vision of hell:

> The whole length of the street, deep as a well and narrow like a corridor, was full of a sombre and ceaseless stir. Our ears were filled by a headlong shuffle and beat of rapid footsteps and an underlying rumour – a rumour vast, faint, pulsating, as of panting breaths, of gasping voices. Innumerable eyes stared straight in front, feet moved hurriedly, blank faces flowed, arms swung. Over all, a narrow ragged strip of smoky sky wound about between the high roofs, extended and motionless, like a soiled streamer flying above the rout of a mob.

Three times Jackson looks about meditatively at the nightmare city, so full of energy and yet of distress and poverty; and the final image is of a policeman, helmeted and dark, stretching out a rigid arm at the crossing of the streets. This urban force, too, is seen as a living power, the composite machine of commerce and imperialism in opposition to the savage nature which Karain signified. '"Yes, I see it," said Jackson, slowly. "It is there; it pants, it runs, it rolls; it is strong and alive; it would smash you if you didn't look out; but I'll be hanged if it is yet as real to me . . . as the other thing . . . say, Karain's story."' Conrad gives the narrator the last word: the man has been decidedly too long away from home.

Conrad used his experience of the Malay Archipelago to question Western values, creating an ambivalent, shifting set of tensions between the attractions of tropical nature and a way of life adapted to the green environment, and the more measured, guilt-ridden certainties of Western industrial civilisation. But from the same period as 'Karain' emerged another story with a very different feel, 'An Outpost of Progress' – originally entitled 'A Victim of Progress'. This stemmed, like its successor *Heart of Darkness*, from Conrad's four months on the Congo, as first mate and temporary master of the *Roi des Belges*, a true posting from hell. Conrad should never have gone; he was lucky to come back as soon as he did, and even then spent a long time convalescing. 'Everything here is repellent to me. Men and things, but men above all,' he wrote; and the men were either 'black savages' or 'white slaves (of whom I am one)'. He experienced four bouts of fever in two months, and a severe attack of dysentery; and, besides sensing immediately that the whole enterprise was rotten, he felt himself to be on the receiving end of bitter professional hostility from his Belgian superior towards an 'Englishman'. This hostility actually saved him from a trip up the Kassai, which he chose as the setting of this first African tale: a story which he was, he admitted, pleased with. 'All the bitterness of those days, all my puzzled wonder as to the meaning of all I saw – all my indignation at masquerading philanthropy – have been with me once again while I write.' The corruption spawned by the wholesale commercial and colonial exploitation was absolute; the infection spread from the Europeans to the Africans, leaving no one whole; and the sense of tropical Africa as something negative, or at best a landscape in which the negative flourishes, makes an uncomfortable statement. This story, and *Heart of Darkness*, were, he commented, 'all the spoil I brought out from the centre of Africa, where, really, I had no sort of business'.

'The Outpost of Progress' is a heavily ironic reversal of the prototype 'white man into the unknown' narrative, in which some heroic representative of European civilisation upholds standards of decency and Christianity in the heart of the jungle/desert/wilderness. Two uneasily matched agents, refugees from failed lives in Europe, are offloaded, like bales of cargo, at a remote station in the Congo basin, and left for six months while their director heads off upstream by steamer, with little expectation of their success. They are the complete opposite of the 'noble' traders praised by Mary Kingsley. Slowly, they atrophy, trading in a desultory way for ivory, an activity that is a symbol of the destruction of nature, though all transactions are carried out by their Sierra Leonean assistant, suggestively named Price, but called by the local people Makola. Their one positive contact is with the headman of the local village, who sustains them with fowls and corn in exchange for the sense of protection which the outpost affords. Trade is thin; a group from another tribe arrives, behaving, and talking, in a different way: as the two men remark, the rhythms and sounds remind them of French. Makola takes charge. The Europeans retire to their house. A party is proposed for the new group, the station servants, and some men from the village. During the night a shot is heard. In the morning the group has gone, leaving behind them a pile of ivory; but the outpost servants, and the villagers, have vanished, and it becomes clear that the coastal tribesmen were slave traders. Compromised, the Europeans disintegrate, physically and morally; they quarrel, absurdly, over a bag of sugar as their supplies dwindle. One shoots the other; the second hangs himself. When Makola rings the bell in answer to the steamer's hooter, the director comes ashore through the symbolic fog to be confronted by the body of the second European, hanging from a ramshackle cross, which itself marks the grave of the former outpost agent, a failed artist. The corpse's swollen tongue protrudes towards him.

The story appeared in two successive numbers of *Cosmopolis* in 1897, next to an article praising in extravagant terms the achievements of Queen Victoria and her Empire. Simpler in structure and method than *Heart of Darkness*, it similarly explores the way in which European values are exposed as hollow and destructive, when removed from their context and confronted with the impenetrable mystery of Africa, and the tropical forest. But Conrad does not indicate any positive, or even neutral, values and images to set against his devastating critique of the decay of Europe. The forest is seen as menacing; the river, though a source of fish and hippopotamus meat, is little more than a supply route, on which the company's tin steamer may eventually appear out of the fog. The villagers practise their rites; the drums beat in the night; but the language overlays every aspect of life with a message of judgement. Makola/Price, sharing an African and English name, married with a plump wife and children, is the most positive figure; but even he is portrayed principally as a survivor, an uneasy negotiator between two cultures and systems. The messages conveyed by the story are of the destruction and despoliation of Africa by a capitalist Europe, a process which corrupts and destroys the weak of both continents; but the vision is, fundamentally, one of hell. The Europeans travel up the river, as in Conrad's other story, into the heart of darkness, their own darkness, but also an objective darkness.

> Going up that river was like travelling back to the earliest beginnings of the world, when vegetation rioted on the earth and the big trees were kings. An empty stream, a great silence, an impenetrable forest. The air was warm, thick, heavy, sluggish. There was no joy in the brilliance of sunshine. . . . There were moments when one's past came back to one, as it will sometimes when you have not a moment to spare to yourself; but it came in the shape of an unrestful and noisy dream, remembered with wonder amongst the overwhelming realities of this strange world of plants, and water, and silence. And this stillness of life did not in the least resemble a peace. It was the stillness of an implacable force brooding over an inscrutable intention.

In *Heart of Darkness* Marlow's description of his voyage to find Kurtz records a journey into prehistory: 'We were wanderers on prehistoric earth, an earth that wore the aspect of an unknown planet. We could have fancied ourselves the first of men taking possession of an accursed inheritance, to be subdued at the cost of profound anguish and of excessive toil.' When Marlow's narrative comes to an end, and Conrad returns us to the cruising yawl on the Thames, the listener raises his head: 'The offing was barred by a black bank of clouds, and the tranquil waterway leading to the uttermost ends of the earth flowed sombre under an overcast sky – seemed to lead into the heart of an immense darkness.' In Sopot, on the Baltic, a giant statue of Conrad stands at the end of the pier, looking out to sea, as if to his future; but the vision he brought back from Africa was sombre and disturbing. It was essentially an expedition to the past, a journey back, a regression. Kurtz's message to the world was: 'The horror!'

Parallel to Conrad's own journey into the interior, another imaginative source for *Heart of Darkness* was Stanley's 1890 account of his expedition to find Emin Pasha, and free him from the grip of the Mahdists. Stanley's title, *In Darkest Africa*, signals the perspective; and as a journey this was, by any standards, horrible and horrifying, largely because of the suffering everyone experienced from disease and hunger. (Frustratingly Emin Pasha, a scientific traveller naturalised in Africa, did not want to be rescued any more than Kurtz did.) Conrad's nightmare personal experience of the Belgian Congo resonates through *Heart of Darkness*. In spite of the fierce irony, the sense of an imperial role dominates, as though it is inevitable that European races must impose their will on Africa, for all the wrongs that accompany such 'progress', and for all the hypocrisy about motives which sought to conceal the greed and lust for power that drove it forward. There is little sense of the wonder or beauty of the continent, of 'the elemental force of ardent life' which Conrad expressed in 'Karain'. It is not surprising that African writers such as Chinua Achebe have offered a devastating critique of Conrad.

Heart of Darkness is not simply an exposure of commerce and imperialism. It is also a metaphor of the search for man's origins: the journey into the past, into a more primitive state, as well as into the geographical centre of the continent. Instinctively, writers both as subtle as Conrad and as relatively straightforward as Rider Haggard used central Africa as the destination for Europe's search: the throbbing drums, the primeval forest, the idea of a lost kingdom, all speak of an instinctive placing of man's roots in the centre of the most resistant environment on earth. Africa presented a perpetual challenge, to the imagination as much as to the colonist, the missionary or the explorer. Africa resisted vigorously: its people had been there a long time, had learned resilience, and adaptation. Besides, everyone else might have come from there too, in prehistory, however much they had adapted in later periods to new environments, even if this 'Out of Africa' theory was only glimpsed as imagination and speculation, rather than supported by the fossil discoveries of south and east Africa.

The image which emerges from *Heart of Darkness* as the most positive and vital is 'the wild and gorgeous apparition of a woman', who confronts the pilgrims from the shore:

> She walked with measured steps, draped in striped and fringed cloths, treading the earth proudly, with a slight jingle and flash of barbarous ornaments. She carried her head high; her hair was done in the shape of a helmet; she had brass leggings to the knee, brass wire gauntlets to the elbow, a crimson spot on her tawny cheek, innumerable necklaces of glass beads on her neck; bizarre things, charms, gifts of witch-men, that hung about her, glittered and trembled at every step. She must have had the value of several elephant tusks upon her. She was savage and superb, wild-eyed and magnificent; there was something ominous and stately in her deliberate progress. And in the hush that had fallen suddenly upon the whole sorrowful land, the immense wilderness, the colossal body of the fecund and mysterious life seemed to look at her, pensive, as though it had been looking at the image of its own tenebrous and passionate soul.

The image of Africa, magnificent, regal, non-Christian, fecund, confronts the European traders in their steamer: 'Suddenly she opened her bared arms and threw them up rigid above her head, as though in an uncontrollable desire to touch the sky'; then darkness began to fall, and she walked on, passing into the forest, and disappearing.

The centre of Africa became a favourite setting for the mythical adventures of popular novelists such as Rider Haggard and John Buchan. Rider Haggard's African novels have the structure of romance quests: a group of miscellaneous Englishmen plunge into the interior, up crocodile-infested rivers, across deserts, mountain ranges, swamps and forests, heading for a lost kingdom: King Solomon's Mines, Kor, Zu-Vendis. The style and spirit of the stories is action-packed; but they are clearly related to the journals of the explorers and scientific travellers. *Allan Quatermain*, for example, is presented as the last adventure of the great white hunter, with a final chapter 'By Another Hand', and the whole story is legitimised by Sir Henry Curtis's brother, George, who has received the manuscript with an Aden postmark: to add authenticity, there are occasional editorial notes, relating the fabulous action to real events. ('By a sad coincidence, since the above was written by Mr Quatermain, the Masai, in April 1886, have massacred a missionary and his wife – Mr and Mrs Houghton – on this very Tana River, and at the spot described.') The trio who successfully reached, and returned from, Kukuanaland and King Solomon's Mines are bored with civilisation and set off again to search for enlightenment: Quatermain, old Africa hand, who has buried his son in a Yorkshire country churchyard and wishes to die as he has lived, among the wild game and the savages; Sir Henry Curtis, straight as a die, the epitome of the strong, handsome, English gentleman – 'a magnificent specimen of the higher type of humanity'; and the comic sidekick, short and stout, Cyclops with his eyeglass, the ex-Royal Naval Captain Good with a roguish penchant for the ladies. These three musketeers are joined by a noble savage, Umslopogaas, and even by an anti-heroic servant, French of course, Alphonse, who cooks divinely and has an enviable ability to survive (it's presumed that Alphonse eventually brings out the manuscript). So much is familiar and recognisable as the stuff of popular romance. More nineteenth-century is Quatermain's explicit weariness with English civilisation: he finds such a very little gulf between the ways of the savage and the ways of 'the children of light':

> I say that as the savage is, so is the white man, only the latter is more inventive, and possesses a faculty of combination; save and except also that the savage, as I have known him, is to a large extent free from the greed of money, which eats like a cancer into the heart of the white man. It is a depressing conclusion, but in all essentials the savage and the child of civilisation are identical. I dare say that the highly civilised lady reading this will smile at an old fool of a hunter's simplicity when she thinks of her black bead-bedecked sister; and so will the superfine cultured idler scientifically eating a dinner at his club, the cost of which would keep a starving family for a week. And yet, my dear young lady, what are those pretty things round your own neck? – they have a strong family resemblance, especially when you wear that *very* low dress, to the savage woman's beads. Your habit of turning round and round to the sound of horns and tom-toms, your fondness for pigments and powders, the way in which you love to subjugate yourself to the rich warrior who has captured you in marriage, and the quickness with which your taste in feathered headdresses varies – all these things suggest touches of kinship; and remember that in the fundamental principles of your nature you are quite identical. As for you, sir, who also laugh, let some man come and strike you in the face whilst you are enjoying that marvellous-looking dish, and we shall soon see how much of the savage is in *you*.

The view Haggard puts forward, through his narrator Quatermain, is one that emphasises the unity of all races: it is also one which praises the virtues of a simpler way of life, before the mercilessness of Political Economy has taken a hold. The search for this land where more natural values can prosper is full of danger. The expedition is harried, and then attacked, by the Masai, and they leave behind a missionary station run on distinctly commercial lines before heading for the interior past the base of Mount Kenya and on towards a mass of snow-tipped mountains. They gradually lose their porters and belongings: one Askari treads on a puff-adder, the second is drowned; their donkeys die of tsetse fly bites; ammunition runs low. Crossing a vast upland lake in a log canoe, they are sucked into an underground river, which takes them down and down, hour after hour: the temperature rises – they strip off their clothes, and are sped past a pillar of fire, to be revived and reborn again on the other side. They find themselves no longer underground but in a chasm, a literal realisation of Coleridge's sacred river, Alph. There they sit on the rocks to recover, and are attacked by a huge species of black, freshwater crab. When Umslopogaas cracks one open with his axe, a terrible screaming echoes round the cavern, a fearful stench is released, and the whole scene becomes like an image of hell from Dante's *Inferno*. As Quatermain comments, there was something unnerving about the encounter: 'there was something so shockingly human about these fiendish creatures – it was as though all the most evil passions and desires of man had entered the shell of a magnified crab and gone mad. They were so dreadfully courageous and intelligent, and they looked as if they understood.' After this last encounter with the old world, and with a primeval form of life, the travellers go underground once more, to emerge upon a vast lake, and finally arrive before Milosis, the Frowning City.

As they near the shore, the group dress themselves – Good in the full-dress uniform of a Commander of the Royal Navy (Umslopogaas comments: 'I always thought thee an ugly little man, and fat – fat as the cows at calving time; and now art thou like a blue jay when he spreads his tail out.') As a further gesture to impress the natives who sail out to meet them, armed only with swords, Captain Good, in true imperial style, suggests that they blast off at a school of hippopotami. Good misses, a bull hippopotamus attacks a canoe and almost kills a girl. It turns out that this is a sacred herd, and they are only saved from the death-penalty by the intervention of the queen.

The lost kingdom which the oddly assorted group has reached is a pre-industrial, agricultural country, whose people worship the Sun and are ruled by twin queens, one fair, the elder, Nyleptha;

one dark, Sorais. There is more than a touch of male fantasy about the treatment of the women, accompanied by a half-ironic, half-solemn fear: '"Oh, my word!" thought I to myself, "the ladies have come on the stage, and now we may look to the plot to develop itself." And I sighed and shook my head, knowing that the beauty of a woman is like the beauty of the lightning – a destructive thing and a cause of desolation.' At the same time, there is a hint of eugenics. The two queens sweep into a hall, each the epitome of regal beauty, and their eyes pass over each traveller and rest on Sir Henry Curtis, with the sunlight from a window playing upon his yellow hair and peaked beard: 'and thus for the first time the goodliest man and woman that it has ever been my lot to see looked upon one another'. After an epic struggle, the forces of Nyleptha triumph, and Sir Henry Curtis becomes her consort. Quatermain and Umslopogaas die, wounded and exhausted. Good survives to build a navy on the lake. Curtis remains, committed to the total exclusion of all foreigners from Zu-Vendis, in order to preserve for the 'on the whole, upright and generous-hearted people' the blessings of comparative barbarism: a life without gunpowder, steam, daily newspapers – and, he adds, universal suffrage: 'I have no fancy for handing over this beautiful country to be torn and fought for by speculators, tourists, politicians, and teachers' – and he evokes the image of the crabs in the valley of the underground river. He will not endow Zu-Vendis with the general marks of 'civilisation', the usual legacy of the West: 'greed, drunkenness, new diseases, gunpowder, and general demoralisation'.

Haggard achieves at a stroke the ideal of the imperialist scientist/explorer: he annexes the lost country for the West, in the person and genes of Sir Henry Curtis, and offers the vision of a white (or whitish – as white 'as Spaniards or Italians') near-paradise in the heart of the oldest continent. Curtis, a Viking version of Umslopogaas, is going to sort out the warring factions within Zu-Vendis; and, as a kind of sop to his readership, Haggard credits him with the aim of introducing true religion in the place of sun worship; but the political economy seems more likely to be based on Malthusian principles – the natural law that prevents any animals increasing beyond the capacity of the country they inhabit to support them.

Haggard followed *Allan Quatermain* with another broadly parallel myth *She*, which is more clearly a journey into the past, a search for the origins of civilisation in the heart of Africa. This time, the questors have a more personal mission. They comprise Leo Vincey, known as 'the Greek god'; his guardian, Horace Holly, a Cambridge fellow and mathematical scholar, as ugly as

Vincey is handsome; and Job, a college servant. The Cambridge setting supplies a kind of anchorage to the fantastic journey back in time, to the ancient, pre-Egyptian civilisation of Kor, and the mystery of the Arab semi-divine Queen Ayesha, She-who-must-be-obeyed. In the story's scheme, Vincey is the descendant, or reincarnation, of Kallikrates, slain by Ayesha 2,000 years before; but the most powerful element is the yearning of Holly for the love of Ayesha, of the beast for beauty. Holly is described as 'short, rather bow-legged, very deep chested, and with unusually long arms. He had dark hair and small eyes, and the hair grew down on his forehead, and his whiskers grew quite up to his hair. . . . Altogether he reminded me forcibly of a gorilla.' He thinks of himself as abnormally ugly – once he overheard a woman refer to him as a 'monster' and say that he had converted her 'to the monkey theory'. Billali, Holly's African mentor, calls him 'my son the Baboon'.

The story's climax comes in the Temple of Truth. Greiffenhagen, Haggard's illustrator, juxtaposes a lightly clad Ayesha with two uniformed Victorian explorers, still equipped with puttees, pith helmets and revolvers. When Ayesha prepares to purify herself in fire, so that she can join Vincey, she unveils herself, like Salome before Herod: but then, like Dorian Gray's portrait, her form grows old before their eyes. 'She's shrivelling up! she's turning into a monkey!' 'She, who but two minutes gone had gazed upon us – the loveliest, noblest, most splendid woman the world has ever seen – she lay still before us, near the masses of her own dark hair, no larger than a big ape, and hideous – ah, too hideous for words!' Ayesha has turned into Lucy; and the Victorian scholar's hope of finding new life in the form of beauty mysteriously preserved from a pre-Christian age remains only a dream. Although Ayesha is given authority and choice, the structure of the fable suggests the destruction of natural beauty by the invading English academics, who are, in reality, more akin to Lion and Baboon, as Billali names them. Whether in the heart of Africa or in the streets of London, the transformation of Dr Jekyll into Hyde was a motif which haunted the Victorian imagination, raising the fascinating but appalling spectre of brutalisation.

If John Buchan had been born a century earlier, one could imagine him as an African explorer in the tradition of that earlier Border Scot, Mungo Park, whom he once quoted: 'I would rather go back to Africa than practise again in Peebles.' Buchan went to South Africa in the aftermath of the conflict with the Boers, as one of Milner's young men, his 'kindergarten'. His job was to help get

35. Ayesha – 'She' – offering herself as a sacrifice to Leo Vincey: 'Strike, and strike home!' (Maurice Greiffenhagen's illustration from the 1896 edition of *She*)

the country going again. In the two years he spent there, the idea of Africa imprinted itself on him, and he passed a further two years back in London waiting for a suitable post in Egypt, which never materialised. One of the four longer treks he made was to Wood Bush, in the Zoutpansberg Mountains:

> I have never been in such an earthly paradise in my life. You mount up tiers of mountain ridge, barren stony places, and then suddenly come on a country like Glenholm. Terrific blue mountains rise to the South but the country is chiefly little wooded knolls, with exquisite green valleys between. The whole place looks like a colossal nobleman's park laid out by some famous landscape gardener, and when you examine it closely you find it richer than anything you can imagine. The woods are virgin forest – full of superb orchids and fern, and monkeys, and wild pig, and tiger-cats and bush-buck. The valleys have full clear streams flowing down them and water-meadows – where in place of meadowsweet and buttercup you have tall blue agapanthus and huge geraniums and great beds of arums and the ferns. The perfume of the place is beyond description. The soil is very rich: the climate misty and invigorating, just like Scotland.

Buchan is writing within the tradition of the scientific explorers, marvelling and at the same time appropriating: the place is Scotland without the disadvantages, a kind of tropical deer forest, and he dreamed of buying a small farm there. The Wood Bush stayed in Buchan's mind as one of his series of enchanted places, and he used it as the natural setting for his symposium novel, *A Lodge in the Wilderness*, which takes place in Sir Francis Carey's house at Musuru to the west of the Rift Valley: the figure of Carey is based on Cecil Rhodes, and in the dream country-house are assembled a set of characters who share aspects of Milner, Rosebery, Balfour, Alfred Beit, Lady Lyttelton, Lady Lugard; and a young man, Hugh Somerville, who is, clearly, John Buchan. Each evening they discuss imperialism, and the final definition which emerges is a justification for the scramble for Africa, a justification firmly rooted in the idea of the English as a naturally dominant and superior species, a successful variety:

> England has completed her great era of expansion. Her work for ages was to find new outlets for the vigour of her sons, and to occupy the waste or derelict places of the earth. Now, the land being won, it is her task to develop the wilds, to unite the scattered settlements, and to bring the whole within the influence

of her tradition and faith. This labour we call empire-building, and above all things it is a labour of peace.

In the space of a century, the European perspective on Africa has been transformed, so that Buchan's characters, echoing the political views of the British cabinet, can view the whole continent, indeed the whole world, as an appropriate arena to civilise and, as a necessary corollary, to convert, for Buchan was a son of the manse. This is very much a post-Darwinian apologia, with the white man leading the less advanced in a peaceful march of progress: the conquest of the fittest is over, and the savage will be shown the benefits of civilisation. The notion of exploitation is hinted at only in that harmless-sounding word 'develop', a term which has acquired much more ruthless connotations a hundred years later.

The same 'civilising' impulse is present in Buchan's other African novel, *Prester John* (1910), in which the Reverend John Laputa, who believes himself to be the reincarnated spirit of the legendary African king, plans to lead a revolution against the whites. The dying Laputa hurls himself into an underground chasm; David Crawfurd and Arcoll persuade his followers to lay down their arms; and with the source of diamonds found in the cave, a native college is founded, not to turn out missionaries and teachers, 'but an institution for giving the Kaffirs the kind of training which fits them to be good citizens of the state. There you will find every kind of technical workshop, and the finest experimental farms'. But although the narrative disposes of Laputa and thwarts the planned uprising, the figure which emerges most memorably from the novel is that of Laputa himself, the image of a king in waiting. As with Conrad's forest queen in *Heart of Darkness*, the vision and the power belong to Africa.

Before Conrad wrote about his African experiences, Robert Louis Stevenson had explored some of the same themes in his stories of the South Pacific. But first he probed the idea of civilisation, and the uncomfortable relationship between man and beast, in his rewriting of *Frankenstein*, 'The Strange Case of Dr Jekyll and Mr Hyde'. The story, which Stevenson said came to him in a dream, is placed on the chilling frontier between civilisation and savagery, a location made all the more disturbing because of Stevenson's evocation of the nineteenth-century city, shrouded in fog (London, though it seems more like Edinburgh in places) and of the stuffy respectability of upper-middle-class, professional 'society' – a society, it seems, exclusively made up of wealthy men and servants. Stevenson probes the dual nature of man: on the one hand, the apparently upright, moral Jekyll, who, by drinking a chemical substance, has discovered that he can be transformed at night into the amoral Hyde. (Stevenson makes his Victorian Faust a scientist, though not a very convincing one, reverting to a Calvinistic concept of experiment as forbidden, in contrast to the positive image of Faust which Lessing promoted.) Under cover of darkness, Jekyll as Hyde enjoys a second life: he feels younger, lighter, happier in body: 'within I was conscious of a heady recklessness, a current of disordered sensual images running like a mill race in my fancy, a solution of the bonds of obligation, an unknown but not an innocent freedom of the soul'. Jekyll is not the same man physically as Hyde: Hyde is smaller, uglier – and yet when he looks in a glass he is not conscious of any repugnance, but feels a leap of welcome, for the image seemed natural and human.

The language which Stevenson draws on to describe Hyde is that of the animal, and specifically of the ape. Hyde has no moral sense, no conscience: he drinks pleasure 'with bestial avidity' from one degree of torture to another. A maid chanced to observe Hyde clubbing Sir Danvers Carew to death; 'and next moment, with ape-like fury, he was trampling his victim under foot'. 'Ape-like' is a phrase Jekyll uses of himself; and, what is truly alarming, while he sits in the sun on a bench as Jekyll, the animal within him 'licking the chops of memory', he finds himself changing physically, but this time spontaneously, without recourse to the 'tincture'. 'I looked down; my clothes hung formlessly on my shrunken limbs; the hand that lay on my knee was corded and hairy.' The beast, the ape in man, has taken over: the 'I' who is Jekyll is losing control, and the brute nature which he no longer recognises as his has become dominant. As Conrad's Kurtz does in the heart of Africa, so Stevenson's Jekyll expresses in the streets and alleys of a northern, fogged and benighted city a sense of horror at the thinness of the veneer which separates man from beast. This is humanity doubting any concept of progress and development; there is no notion of ladders here, but a swift and sensational descent down the snakes to the jungle floor.

Stevenson, brought up in the moral rectitude of Calvinist Scotland, left England in 1887, and his search for freedom and better health took him finally to Samoa, about as far away as you can get, physically and temperamentally, from nineteenth-century Edinburgh. The Pacific islands were, in the European imagination, the gardens of paradise, a perfect setting for the dreams of a lotus-eater's existence, often heavily embroidered with male sexual fantasies. Captain Cook may have been killed in a scuffle on the

Hawaiian Islands, but that was a relatively isolated incident. After the mutiny on the *Bounty*, the crew made for Pitcairn Island: a genetically disastrous decision, in the long term, but in that context infinitely preferable to a long voyage under Bligh's naval discipline. By the time Darwin visited Tahiti, the better effects of the missionary movement struck him immediately. The island was like a beautiful orchard of tropical plants, and the people showed an intelligence 'which shows they are advancing in civilisation'; as he put it to his sister, 'The kind simple manners of the half civilized natives are in harmony with the wild, & beautiful scenery.' Here, Darwin was confident, the native peoples could only benefit from the civilising, philanthropic efforts of the colonisers.

At the end of the century, and freed by distance from the values of Europe, Stevenson presented a very different scenario. His feelings were complex, as he explores them both in his fiction and his travel book, *In the South Seas*. 'Few men who come to the islands leave them. The first experience can never be repeated. The first love, the first sunrise, the first South Sea island, are memories apart and touched by a virginity of sense.' But the Marquesans disturbed him at first, squatting cross-legged on the cabin floor, regarding him in silence with embarrassing eyes. He had come to a different world: 'I was now escaped out of the shadow of the Roman empire, under whose toppling monuments we were all cradled, whose laws and letters are on every hand of us, constraining and preventing. I was now to see what men might be whose fathers had never studied Virgil, never been conquered by Caesar.' He wondered if his subsequent friend, Kauanui, 'might leap from his hams with an ear-splitting signal, the ship be carried at a rush, and the ship's company butchered for the table'. Stevenson comments, 'There could be nothing more natural than these apprehensions, nor anything more groundless.'

As Stevenson made his way through the South Seas he transferred his immediate empathy for the people and their way of life into active campaigning, writing letters, for example, to the London *Times* on the political situation in Samoa, which had been subjected to the same kind of carve-up as Africa in the June 1889 Treaty of Berlin. When armed rebellion broke out in 1893, Stevenson organised a field hospital in Apia and nursed the wounded supporters of Mataafa. When they were freed after an amnesty, some of them constructed a road through the forest to Stevenson's home at Vailima: 'The Road of the Loving Heart'. Following his heart, Stevenson constructed an alternative interpretation of the people and their precarious, finely balanced society: which abruptly challenged the superiority of the West, and the

West's values, being ruthlessly imposed on the fragile system that is evoked in Stevenson's description of a village:

> It was longer ere we spied the native village, standing (in the universal fashion) close upon a curve of beach, close under a grove of palms; the sea in front growling and whitening on a concave arc of reef. For the cocoa-tree and the island man are both lovers and neighbours of the surf. 'The coral waxes, the palm grows, but man departs,' says the sad Tahitian proverb; but they are all three, so long as they endure, co-haunters of the beach.

'The Beach of Falesá' explores the interaction of two cultures, through a 'poor white' trader, Wiltshire, who is landed on the beach to take over an isolated station, and trade in copra and exploit the natives, in a situation comparable to that of Kurtz, on the Congo. Wiltshire is taken in hand by a suspiciously friendly degenerate trader, Case, who says he must help him to a wife. Wiltshire agrees – it's the normal thing:

> There was a crowd of girls about us, and I pulled myself up and looked among them like a bashaw. They were all dressed out for the sake of the ship being in; and the women of Falesá are a handsome lot to see. If they have a fault, they are a trifle broad in the beam, and I was just thinking so when Case touched me.
>
> 'That's pretty,' says he.
>
> I saw one coming on the other side alone. She had been fishing; all she wore was a chemise, and it was wetted through. She was young and very slender for an island maid, with a long face, a high forehead, and a shy, strange, blindish look between a cat's and a baby's.
>
> 'Who's she?' said I. 'She'll do.'

So far, so relatively conventional: the white man's sexual exploitation is presented as a matter of course, almost a fact of economic expectation on the part of the island girls. By dusk, Case has concluded negotiations with Uma's mother, and she has come for a 'marriage' ceremony, conducted by Case's Negro friend dressed in a big paper collar, who read words of a service 'not fit to be set down' from an odd volume of a novel. Case writes out the marriage certificate:

> This is to certify that *Uma*, daughter of *Faavoo*, of Falesá, island of —, is illegally married to *Mr John Wiltshire* for one night, and Mr John Wiltshire is at liberty to send her to hell next morning.
>
> John Blackamoor,
> Chaplain to the Hulks

36. The marriage scene from 'A South Sea Bridal', chapter 1 of 'The Beach of Falesá'. (*Illustrated London News*, 9 July 1892)

(This was too much for the editor of the *Illustrated London News*, who removed the infamous certificate altogether when he printed the story, along with some of the stronger language. Even when it was reprinted, 'one week' was substituted for 'one night', as if that made things much better.)

It slowly transpires that the 'marriage' to Uma has been engineered by Case. Uma is tabooed, and so too now, by his association with her, is Wiltshire. No one will trade with him, and Case is able to keep the monopoly. But, stage by stage, a change comes over Wiltshire. He grows to – well, love is not a word in his vocabulary, but by this time she has become his 'old lady'. When he learns the truth, he acts on instinct. A missionary arrives on one of his intermittent visits: Wiltshire asks him to perform a proper marriage ceremony. Then he goes after Case. Case has constructed a kind of devil-area in the island, to frighten the wits out of the natives – with a Tyrolean harp, luminous paint and various other devices, a kind of open-air ghost-train. Wiltshire lies in wait for him there, accompanied by Uma, and finally kills him in a fight to the death. Tarleton, the missionary, puts in a good word for Wiltshire, and the natives begin to trade with him – though, much to Wiltshire's annoyance, the missionary tells him he has given his pledge that Wiltshire will deal fairly. It comes as a considerable relief when he is moved to another island and can resume trading in the 'normal' way. As for Uma, she's an A-1 wife, though 'if you don't keep your eye lifting she would give away the roof off the station. Well, it seems it's natural in Kanakas.'

Stevenson has chosen as his central character, and narrating voice, an uneducated, hard-headed Englishman, who has all the self-centredness and bigotry of his age, but who has a reservoir of morality which has responded to the 'natural' goodness of Uma. Wiltshire has given up his former dream, to earn enough money to buy a public house in England. He has a wife and children, and his actions have left him as isolated, socially, in his own eyes as he was when he was tabooed. Stevenson, without in any way attempting to suggest an ethical solution to such a tricky subject, leaves the narrative with the image, and dilemma, of racial interbreeding, like a problem tossed towards the next generation, and the next century, to solve. Appropriately, he signs off with a question:

> I don't like to leave the kids, you see: and – there's no use talking – they're better here than what they would be in a white man's country, though Ben took the eldest up to Auckland, where he's being schooled with the best. But what bothers me is the girls. They're only half-castes, of course; I know that as well as you do, and there's nobody thinks less of half-castes than I do; but they're mine, and about all I've got. I can't reconcile my mind to their taking up with Kanakas, and I'd like to know where I'm to find the whites?

Stevenson, weary of the politeness and hypocrisy of white society, was bold enough to confront the most volatile question highlighted by the nineteenth-century scientists: how would the human race evolve? There were two possible and opposing ways, two interpretations of the survival of the fittest, even if the mechanism of inheritance, the genetic system, was not yet understood. The two images of Conrad and Stevenson, the African queen and the island girl, suggest the choice. In *Heart of Darkness*, the white pilgrims are confronted by the magnificent woman standing on the river shore, stretching out her arms to the sky, while Kurtz, her potential partner, lies hidden from view; and, when the steamer heads downstream, Marlow pulls the string of the whistle to scatter the mass of naked, breathing, quivering, bronzed bodies.

> Only the barbarous and superb woman did not so much as flinch, and stretched tragically her bare arms after us over the sombre and glittering river. And then that imbecile crowd down on the deck started their little fun, and I could see nothing more for smoke.

The Europeans' rifles mow down the Africans; the 'savage' race is slaughtered by the 'civilised'; and the figure of the potential Eve, the counterpart to Kurtz's pale, fair-haired Intended, is killed. This is one response to the new knowledge, which fits very neatly into the imperial and commercial instincts of late nineteenth-century Europe. Francis Galton, Darwin's brilliant and eccentric cousin, had no doubt that eugenics was a science. If farmers could breed more effective domestic animals, then it should be possible to improve the human race – which, by and large, meant the white race, and the more advanced and intelligent and healthy specimens among those: people like Galton. By the same kind of thinking, the struggle for survival would lead to the elimination of 'inferior' races. The whole scramble for Africa can be seen as an exercise in social Darwinism of the most ruthless, international kind. The empire-building Harry Johnston's argument was representative and influential: 'The Negro seems to require the intervention of some superior race before he can be roused to any definite advance from the low stage of human development in which he has contentedly remained for many thousand years.'

The opposing view was put more cautiously by Stevenson. Uma is first depicted by male, European eyes, an island maid, an innocent Eve; but, as the story unfolds, the complexity of her past, of her consciousness and moral feelings, begins to be suggested, until the sense of integrity, of moral evolution, is centred on her. Wiltshire recognises this, though he cannot articulate it. Their partnership, their survival, their children, all point to a potential breeding out of the baser kind of attitudes, represented by Case, and by the Wiltshire who first landed at Falesá. The instinctive generosity and goodness in a society not dominated by property is

37. Uma. (*Illustrated London News*, 9 July 1892)

conveyed without sentiment by Wiltshire's comment: 'It seems it's natural in Kanakas'; and Stevenson would have known that 'Kanaka' meant 'man'. The implications of Stevenson's ideas were already present in the thinking of Wallace, as he speculated on the bearing natural selection may have for the human race:

> If my conclusions are just, it must inevitably follow that the higher – the more intellectual and moral – must displace the lower and more degraded races; and the power of 'natural selection', still acting on his mental organization, must ever lead to the more perfect adaptation of man's higher faculties to the conditions of surrounding nature, and to the exigencies of the social state. While his external form will probably ever remain unchanged, except in the development of that perfect beauty which results from a healthy and well organized body . . . his mental constitution may continue to advance and improve till the world is again inhabited by a single homogeneous race.

Wallace is careful here not to specify which races are 'the more intellectual and moral'; Stevenson offers a direct challenge to the accepted definitions. The pictures in the *Illustrated London News* which accompanied serial publication of 'The Beach of Falesá' are very clear. The 'marriage' ceremony condemns the old debased world; the natural figure of Uma, framed by palm-fronds, is a new Eve, not so much a romantic image of a possible paradise as a warning about the fragility of life.

10 *Through the Looking-Glass*

THE IDEAS OF NATURAL SELECTION and the survival of the fittest provided the key to a new, radical and, to many, alarming view of the world. Darwin articulated the theory, and did it in a book (unreferenced, scarcely illustrated) which was accessible to non-scientists. The view still causes alarm; and the ideas which sustain it have undergone modification and refinement in the twentieth century. But it still seems the single most revolutionary rewriting of humanity's relationship to the world in modern times, and has taken a century or more to digest. (It was possible in the 1950s, if you were not a scientist, to be considered sufficiently educated to enter university and yet to know nothing material about natural selection.) These laws were the new tablets of stone, brought to the waiting people not miraculously from the mountain, but fragment by fragment, specimen by specimen, from the heart of the equatorial forests and the depths of the oceans. Ironically, it needed the Industrial Revolution to make the task possible, to rediscover and rethink the natural world, because it was only the industrialised, civilised races who had the pressing impulse to classify and name on a world scale, having prised themselves free from more contained ecosystems.

The implications of natural selection were and are shocking. First, there is nothing special about *Homo sapiens*, apart from the fact (an admittedly massive qualification) that we think about ourselves in a way that seems to be unique. Huxley would joke about his mass audiences: 'By next Friday evening they will all be convinced that they are monkeys.' It was probably better to make a joke about the new truth, as it required a massive intellectual repositioning to come to terms with so radical a definition. The spectre of brutalisation, bestialisation, transmutation, haunted the English imagination: was the hairy Hyde lurking beneath the skin of each Dr Jekyll? The distance and contrast between an English male specimen of the second half of the nineteenth century – to take some unfair examples, a captain of industry, or a peer, or a general – and a naked ape was apparently immense. There was so much surface clutter to begin with, all those layers of clothes, and sticks and gloves, and monocles and cigar cases and cigar cutters, to be disposed of, before coming to the social and cultural attitudes, let alone the ideas and beliefs, which had to be confronted. When some young English settlers reached Port Lyttelton on the South Island of New Zealand in the 1850s, they were so delighted at the prospect of freedom that, on their first evening ashore, they built a huge bonfire, piled their top hats and tail-coats on it, and danced in a ring round the blazing fire: a first step towards becoming more natural, but one which could only safely be taken after sailing to the other side of the world. In Thoreau's words, 'Sell your clothes and keep your thoughts.'

A number of fundamental shared social beliefs have changed since pre-Darwin times – it might be more relevant to say pre-Wallace, since the human dimension was so often at the forefront of his thinking. *Homo sapiens* is no longer such a distinct concept, set apart, specially created by divine intervention. We (a dangerous projection, but it seems less pompous than other evasions) recognise that we are part of a spectrum of species, one that has evolved in a particular way and time to occupy a niche in the environment. We look at the apes – gorilla, chimpanzee, orang-utan – and see creatures with almost as many similarities to ourselves as differences. The distance between ourselves and the other living things has diminished. We share the same space. We share their diseases, their organs; we control, in some measure, their breeding; in the wild, we try in some cases to keep them alive, preserve them from extinction.

We try to keep ourselves (family, friends, neighbours, nationals) alive; and, not very consistently, perhaps, and with opposing outbreaks of genocide or massive indifference, we try to keep other peoples alive: responding to impulses of altruism and co-operation, in an apparent denial of the bald principle of the survival of the fittest, we send food and medicines to the starving and diseased; or at least for a time. We look after the sick, the elderly, help the infertile to have children – or make convincing gestures in those directions.

We have slowly acquired a different attitude towards race. This is a dangerous claim, challenged by almost any edition of any newspaper. You have to be a long-standing optimist to make it, and legislation is needed to back it up. But the myth of race superiority, although it survives in many forms, and not just in white races, has acquired the status of myth. In the nineteenth century it was myth masquerading as fact, myth sanctified by science: brain-pans, eugenics, human family trees with the Europeans/Americans at the apex, who looked down on the lesser breeds as they ruthlessly, behind a smiling mask of benevolence, seized dominion over palm and pine. Now, if anyone is tempted towards a justification of racial superiority, the living memory of the Holocaust intervenes. No one can argue with any chance of being taken seriously outside some minority laager that the white races, for example, have a God-given right to rule: the change in South Africa has closed that chapter. It is interesting to compare what has happened in Africa to the predictions of Sir Harry Johnston at the very end of the nineteenth century. Johnston was a distinguished colonial administrator, and his enthusiasm for empire had been a factor in Lord Salisbury's policy-making. His volume for the Cambridge Historical Series, entitled *A History of the Colonization of Africa by Alien Races*, begins, significantly, with the convenient and wrong theory that the Africans 'in all probability first entered Africa from Asia', thus giving them the status of immigrants. It concludes with a chapter in which the likely pattern of development in a continent, whose map was already coloured red and blue and green, is predicted: 'We have now seen the result of these race movements during three thousand years which have caused nations superior in physical or mental development to the Negro, the Negroid, and the Hamite to move down on Africa as a field for their colonization, cultivation, and commerce.' Johnston foresees that European nations or national types will dominate, with the black population pushed into the unhealthiest regions; but if the other healthy quarters of the globe become over-populated, and science finds an answer to the unhealthy effects of a tropical climate, then a rush might be made by Europeans for settlement 'on the lands of tropical Africa, which in the struggle for existence may sweep away contemptuously the pre-existing rights of inferior races'. The struggle for existence would prevail: this would be social Darwinism on a continental scale. If the struggle has prevailed, then the mechanism has worked in a less predictable and less simplistic way; and if anything has been swept away, or at least undermined, it is the concept of superior and inferior races in its most blatant nineteenth-century form.

One further quote from Johnston emphasises the idea of Africa – though it could equally be India, Ceylon, the Caribbean – as a resource. He openly and unashamedly uses the term 'plantation colonies': 'vast territories to be governed as India is governed, despotically but wisely, and with the first aim of securing a good government and a reasonable degree of civilization to a large population of races inferior to the European. Here, however, the European may come in small numbers with his capital, his energy, and his knowledge to develop a most lucrative commerce, and obtain products necessary to the use of his advanced civilization.' Johnston envisaged an endless scenario in which the raw produce of the rest of the world flowed back to Europe to support a higher level of civilisation, civilisation being another word for consumption. The natural abundance of the tropics was apparently there for the taking. This was John Buchan's philosophy, with teeth.

The process, so devastating to particular places, so amazingly diverse in operation, put into practice on a large scale what the scientific travellers did in miniature, whether it was beetles or butterflies from the Amazon, or cinchona plants from the Andes, or fish from the West African mangrove lagoons, or orang-utans pickled in alcohol from the Borneo forest. Specimens for scientific purposes, or private collectors, or feathers for ladies' bonnets – the dividing lines were blurred. The phrase 'sustainable development' had not been coined, though Wallace observed the way a New Guinea tribe could live in harmony with the environment, feeding and housing and clothing itself in an apparent equilibrium, and pointed up the contrast between a so-called 'savage' community and the raw barbarism of London.

What began to emerge from the collective experience of the scientific travellers was a rediscovered sense of the natural world, and of the interdependence of all forms of life. For individuals to perceive this, it was necessary to leave the centres of primarily urban civilisation, and cross the oceans to live in the tropical forest. The further they moved from England and orthodoxy, the more the young scientists were thrown upon their own intellectual resources as they sought to make sense of undescribed geological formations or unknown forms of life, and the more clearly they seemed to think.

Bates's and Wallace's dreams of living a harmonious, well-supplied, peaceful life in the heart of the Amazon forest gave a fresh impetus to the sense that the natural world could supply everyone's needs. Even for the socialist and humanitarian Wallace, perhaps inevitably, this vision had a colonial dimension – a few well-chosen, like-minded companions would be needed to make the experiment work. Similarly Spruce, pushing up the Amazon tributaries to the Andes, longed for the area to become an English sphere of influence; and Huxley, appalled by the condition of the slaves in Rio de Janeiro, speculated on what the surplus population of England could do to create a better way of life if transported to the region. But Wallace was able to comment from first-hand knowledge, and extended experience, when he wrote: 'The more I see of uncivilized people the better I think of human nature, and the essential differences between civilized and savage men seem to disappear'; and the context for the uncivilised was the natural environment in its most benign form, the forest. Nature was no longer to be seen as something to be tamed, or, alternatively, worshipped at intervals from within some carefully constructed experience, such as Sir Richard Hill's Hawkstone or Girardin's tribute to Rousseau at Ermenonville. Nature was redefined and repositioned as the vital force, a concept greater than any one species, and one which gave life and meaning to everything else: one that needed to be studied minutely; one that, finally, needed respect, protection, preservation.

In 1845 Henry David Thoreau went to live in a hut in the woods overlooking Walden Pond. Huxley was still sweating at his studies at Charing Cross Hospital. Wallace and Bates were beetle-hunting in Leicestershire. Darwin, at Down House, had completed his essay and was pondering the question of transmutation. Thoreau, moved by his mentor Emerson's essay on Nature, went to Walden Pond, as he put it, 'not to live cheaply nor to live dearly there, but to transact some private business with the fewest obstacles': to live, and to think. He did this, largely, on his own, not as part of a community, as in the utopian attempts such as Hawthorne's Brook Farm experiment, or Alcott's Fruitlands. He was not very far from Concord, and he had visitors to talk to, but he was, to all intents, alone with nature; and in this free state he looked at the New England civilisation around him and developed a fresh perspective towards it. Although Thoreau's scale at first seems small, and his concerns domestic, he reaches out in space and time until he is looking at the whole world, as in the conclusion to 'The Bean-Field':

> We are wont to forget that the sun looks on our cultivated fields and on the prairies and forests without distinction. They all reflect and absorb his rays alike, and the former make but a small part of the glorious picture which he beholds in his daily course. In his view the earth is all equally cultivated like a garden. Therefore we should receive the benefit of his light and heat with a corresponding trust and magnanimity. What though I value the seed of these beans, and harvest that in the fall of the year? This broad field which I have looked at so long looks not to me as the principal cultivator, but away from me to influences more genial to it, which water and make it green. These beans have results which are not harvested by me. Do they not grow for woodchucks partly?

Thoreau worked his way outwards from his cabin and the clearing on the edge of Walden Pond, and emphasised the smallness of human beings within the scheme of nature: 'These may be but the spring months in the life of the race.' As he wrote his conclusion to *Walden*, he probed the relationship between 'here' and the rest of the world:

> What does Africa, – what does the West stand for? Is not our own interior white on the chart? black though it may prove, like the coast when discovered. Is it the source of the Nile, or the Niger, or the Mississippi, or a North-West Passage around this continent, that we would find? Are these the problems which most concern mankind? Is Franklin the only man who is lost, that his wife should be so earnest to find him? . . . Be rather the Mungo Park, the Lewis and Clarke and Frobisher, of your own streams and oceans; explore your own higher latitudes, – with shiploads of preserved meats to support you, if they be necessary; and pile the empty cans sky-high for a sign.

Thoreau reminds his readers that 'We are acquainted with a mere pellicle of the globe on which we live', and establishes a sense of scale in an image of himself as he stands over an insect crawling amid the pine needles on the forest floor. Then he shifts the focus to political structures: 'We do not believe that a tide rises and falls behind every man which can float the British Empire like a chip, if he should ever harbor it in his mind. Who knows what sort of seventeen-year locust might next come out of the ground? The government of the world I live in was not framed, like that of Britain, in after-dinner conversations over the wine.'

Thoreau, writing before *The Origin of Species*, anticipates many of the implications of a concept of evolution grounded in natural selection. In his part of New England, the Indians passed by his cabin and through the woods, figures on the periphery of the civilised, over-housed, overdressed society who by contrast helped to suggest to him a more enduring, more natural way of life. Spruce, in the headwaters of the Amazon, could appreciate the interdependence of the true, un-Christianised, Indians with their environment, much as Wallace did in his detailed observations of the Aru islanders off the coast of New Guinea. Wallace foresaw the destruction of one way of life by another, of the East and South by the West and North, of the savage by the civilized; but he also questioned those raw nineteenth-century categories.

Wallace lost himself in the search for new forms of life. On the Aru Islands he was plagued by sand-flies. They attacked his feet, and the resulting ulcers prevented him from walking for a while. He would crawl down to the riverside to bathe, where he might see the blue-winged *Papilio Ulysses*, or some other equally rare and beautiful insect. He could put up with the discomfort and the irritation, but the frustration of being kept a prisoner 'in so rich and unexplored a country, where rare and beautiful creatures are to be met with in every forest ramble' – it was 'a punishment too severe for a naturalist to pass over in silence'. The Aru people could not understand what he was doing with all the animals and birds and insects and shells he was so carefully preserving. Wallace tried to explain that they would be stuffed, and made to look as if they were alive, and that people would go and see them. This did not seem a very convincing explanation, when there would be so many other wonderful things like calico and glass and knives to look at. Eventually, an old man came up with a more satisfactory theory. When the specimens were packed off by sea, they all came to life again. Wallace's specimens, or Marianne North's paintings, brought the natural world to life again, and brought the natural world to the centre of attention.

Wallace, like all the scientific travellers who lived long enough in one place to become more than a fleeting show, passed through the looking-glass, and was able, from time to time, to become an object, and see himself through the eyes of others. He records the same old man of Aru, who bore a 'ludicrous resemblance' to a friend of his in England, indignantly rejecting 'England' as the name of his country, as no one could pronounce it satisfactorily. '"Unglung!" said he, "who ever heard of such a name? ang-lang – anger-lang – that can't be the name of your country; you are playing with us. My country is Wanumbai – any body can say Wanumbai. I'm an 'orang-Wanumbai'; but, N-glung! who ever heard of such a name? Do tell us the real name of your country, and then when you are gone we shall know how to talk about you."' Wallace had given advice on the Crystal Palace exhibition of South American Indians. He had been 'one of the gazers' at the Zulus and the Aztecs in London. In Aru, he realised that he himself was 'a new and strange variety of man, and had the honour of affording to them, in my own person, an attractive exhibition gratis'. By the very act of going to a place such as the Aru Islands, and recording himself there, Wallace helped to create a new perspective on the rest of the world. For a while, he could live like a hunter-gatherer, and propose a vision of the forest as an alternative place to exist in, rather than an environment to be cut down and exported back to 'civilisation'.

The ideas that follow from Darwinian evolution can lead in two directions. The gorilla is either an ugly monster emerging from the primeval darkness, a fearful parody of the human form, or a wonderful species related to ourselves which it is important to study and preserve; and the relationship between us and the gorilla, or the lemur, or the sea squid can be interpreted so as to denigrate man, or to elevate the whole of life. It is the same with the idea of the spirit or the soul, and whether this is compatible with the new science, a problem which particularly exercised Wallace in his attempts to reconcile hard-line Darwinism with a spiritual or moral dimension: the location of the spiritual as part of the mind can lead either to a bleak nihilism or towards an immensely demanding Nietzschean freedom. The naturalists collectively carried out the huge task of supplying the raw data for the scientific revolution; but the very fact of their own worldwide travelling and distribution altered, imperceptibly at first but fundamentally, the European view of the rest of the planet. When Hooker found a specimen of shepherd's purse in the Falklands, and another in the Himalayas, he was filling in a minute corner of a complex, all-embracing system of geographical distribution which connected every part of the earth to every other. When Barth recorded the taxation system of the kingdom of Bornu, or when Mary Kingsley lived alone among the West Africans and saw things worth seeing in what she called their mind-forest, they were beginning the construction of a world which, from a European point of view, was no longer exclusively Eurocentred. The significance of Kew is no longer primarily to serve as the focal point of a network of collectors, who brought specimens back like treasure from the Indies to be preserved by a triumph of science over nature, but as a pivot to turn the mind outwards to the unity and interdependence of the earth's plants.

Chinua Achebe, Africa's responder to Conrad, commented that 'travellers with closed minds can tell us little except about themselves'. For all their inevitable luggage of individual and shared preconceptions and prejudices, travellers like Kingsley and North, Huxley and Hooker, Wallace and Spruce developed amazingly open minds. Whether or not man emerged first in Africa and spread outwards, their collective insights began to reposition Western man within a multi-ethnic order. They drew attention to the connections and relationships between things; recognised that the world was changing, not fixed; emphasised unity, while celebrating variety.

Two contrasting images point up the revolution in thought and attitude. The first is Rousseau's tomb at Ermenonville. Its site is beneath a row of poplars on an island in a lake, the climax of Girardin's carefully landscaped promenade of homage: man placed within a nature sculpted with loving care, essentially controlled and static, a reflective, philosophical and sentimental experience. Rousseau's *Social Contract* provided the intellectual underpinning for the French Revolution; and Girardin's tribute springs from an impulse to create a model for a new paradise in the heart of Europe. The second is Richard Spruce, canoeing down the upper reaches of the Uaupés, the tops of the trees bright with unknown species, and beyond them rich and uncollected swathes of territory. He had bought a telescope at Barra from a Franciscan friar, which enabled him to distinguish green flowers on a tree a mile away. Spruce covered his eyes in frustration so as not to see the beauty he was leaving behind, and the countless species he would never describe. The naturalist here rediscovers a vision of paradise, in a world outside his own.

38. The Red Bird of Paradise (*The Malay Archipelago*).

Bibliographical Notes

INTRODUCTION: TO THE WORLD'S BEGINNING

It will be obvious to anyone who has survived to here that I am greatly indebted to many other writers. In such a general survey it does not seem appropriate to footnote everything, but I hope to make clear what the main sources are, acknowledge the books I have used and referred to, and steer the interested reader towards more detailed commentaries and studies.

Several books have informed me throughout the writing of this exploration, although their authors might not always recognise signs of influence. These include Don Cupitt's *The Sea of Faith* (BBC, 1984) and John Passmore's *The Perfectibility of Man* (Duckworth, 1970). I have read a number of Stephen Jay Gould's essays, notably those in *Ever since Darwin* (Burnett Books/André Deutsch, 1978). At a later stage, I read with great interest Mary Louise Pratt's *Imperial Eyes: Travel Writing and Transculturation* (Routledge, 1992). For earlier travellers in South America, see Anthony Smith, *Explorers of the Amazon* (Viking, 1990). For the social background of the development of natural science, David Elliston Allen's social history, *The Naturalist in Britain* (Allen Lane, 1976), has been invaluable, learned and entertaining. For an account of the writing of 'Kubla Khan', see Richard Holmes, *Coleridge: Early Visions* (Hodder & Stoughton, 1989). Other books which permeate my commentary include Edward W. Said, *Culture and Imperialism* (Chatto & Windus, 1993), Gillian Beer, *Darwin's Plots: Evolutionary Narrative in Darwin, George Eliot and Nineteenth-century Fiction* (Routledge, 1983), John Carroll, *Humanism: The Wreck of Western Culture* (Fontana, 1993), Adrian Desmond, *The Politics of Evolution: Morphology, Medicine, and Reform in Radical London* (U.C.P., 1989) and George W. Stocking, Jr., *Victorian Anthropology* (Macmillan, 1987).

CHAPTER 1: THE SCIENTISTS OF THE SURVEY

The quotations from Darwin's letters are taken from *The*

Correspondence of Charles Darwin, vol. 1, *1821–1836*, edited by Frederick Burckhardt and Sydney Smith (Cambridge University Press, 1985–). I have mostly used the second edition of Darwin's *Journal of Researches*, and have referred also to the abridged text with an introduction by Janet Browne and Michael Neve (Penguin, 1989). On Darwin's life, I am in debt to the great biography by Adrian Desmond and James Moore, *Darwin* (Michael Joseph, 1991). For Hooker, see *Life and Letters of Sir Joseph Dalton Hooker*, 2 vols, edited by Leonard Huxley (John Murray, 1918). On Huxley, see *Life and Letters of Thomas Huxley*, 2 vols, edited by Leonard Huxley (Macmillan, 1900); T.H. Huxley's *Diary of the Voyage of H.M.S. Rattlesnake*, edited by Julian Huxley (Chatto & Windus, 1935), and Adrian Desmond's recent biography, *Huxley: The Devil's Disciple* (Michael Joseph, 1994).

CHAPTER 2: THE HEART OF AFRICA

My *Atlas to Walker's Geography* (London, 1799) cost me five shillings in a second-hand bookshop in Clapham in 1963 (it belonged to a Miss Reynolds – was she a daughter of a member of the Clapham Sect?). I took with me to Nigeria *Africa and the Victorians: The Official Mind of Imperialism*, by Ronald Robinson and John Gallagher with Alice Denny (Macmillan, 1961): it still bears the mould marks of a year in the Niger Delta. See this for the Palmerston quotation. Other books on Africa and African exploration I have referred to include Thomas Pakenham's *The Scramble for Africa* (Weidenfeld & Nicolson, 1991), James Wellard's *The Great Sahara* (E.P. Dutton, 1965) and *Borrioboola-Gha: The Story of Lokoja, The First British Settlement in Nigeria*, by Howard J. Pedraza (Oxford University Press, 1960). Mungo Park's *Travels in the Interior Districts of Africa, 1795–97* first appeared in 1799. Richard Lander's *Records of Captain Clapperton's Last Expedition to Africa*, 2 vols, was published in 1830. Richard and John Lander's *Journal of an Expedition to Explore the Course and Termination of the Niger*, 3 vols, was published by John Murray in 1832. Francis Galton published *The Art of Travel* in 1855 and *Hints for Travellers* in 1878. I have referred to and quoted from his *Memoirs of my Life* (1908) and the 1889 Minerva Library edition of his *Narrative of an Explorer in Tropical South Africa*, first published in 1853. There is an excellent study of Mansfield Parkyns by Duncan Cumming, *The Gentleman Savage* (Century Hutchinson, 1987). Heinrich Barth's *Travels and Discoveries in North and Central Africa* was reprinted in a centennial edition in 1965, with an excellent introduction by A.H.M. Kirk-Greene. For Baikie's letter to Darwin, see Darwin's

Correspondence, vol. 6, *1856-1857*. John Whitford's *Trading Life in Western and Central Africa* was published in Liverpool in 1877.

CHAPTER 3: THE NATURALISTS IN THE AMAZONS

For general background, see again Anthony Smith's *Explorers of the Amazon*. William H. Edwards's *A Voyage up the River Amazon* was published by John Murray in 1847. Charles Waterton's immensely colourful *Wanderings in South America* was first published in 1825. A.R. Wallace's *Travels on the Amazon and Rio Negro* was published in 1853, I have used the slightly revised edition of 1889 (Ward, Lock). H.W. Bates's *The Naturalist on the River Amazons* was published by John Murray in 1863. For Wallace's life, the major sources are A.R. Wallace, *My Life: a Record of Events and Opinions*, 2 vols (Chapman and Hall, 1905) and James Marchant (ed.), *Alfred Russel Wallace: Letters and Reminiscences*, 2 vols (Cassell, 1916). Among more recent studies is Wilma George, *Biologist Philosopher: A Study of the Life and Writings of Alfred Russel Wallace* (Abelard-Schuman, 1964) and H. Clements, *Alfred Russel Wallace: Biologist and Social Reformer* (1983). Other studies include *Wallace and Bates in the Tropics*, edited by Barbara G. Beddall (Macmillan,1969), a comparative selection of their writing.

CHAPTER 4: FROM THE AMAZON TO THE ANDES

The main source for Richard Spruce is his *Notes of a Botanist on the Amazon and Andes*, 2 vols, edited 'and condensed' by his friend Alfred Wallace, and published by Macmillan in 1908. This has been reprinted (Johnson Reprint Corporation, New York) with a foreword by Richard Evans Schultes, who has also published a number of articles on Spruce, among them 'Some Impacts of Spruce's Amazon Explorations on Modern Phytochemical Research', *Rhodora*, vol. 70, 783 (1968). The Linnean Society meeting at Castle Howard in September 1993 disseminated much information about Spruce, and Brian W. Fox produced a paper describing the Spruce material in Manchester Central Library, together with a description of his personal herbarium. The transcript of Spruce's letter to the 6th Countess of Carlisle was kindly supplied by the archivist at Castle Howard. A.S. Byatt's novellas, *Angels and Insects* (Chatto & Windus, 1992), reflect in places the travels and thinking of Wallace, Bates and Spruce.

CHAPTER 5: THE PLANT-HUNTERS

For the general background, D.E. Allen's *The Naturalist in Britain* has been indispensable, and I also referred to Tyler Whittle's *The Plant Hunters* (Heinemann, 1970) and two older classics, F. Kingdon-Ward's *The Romance of Plant Hunting* (1924) and E.H.M. Cox's *Plant Hunting in China* (Collins, 1945). For Paxton and Gibson, see Violet Markham's *Paxton and the Bachelor Duke* (Hodder & Stoughton, 1935). Mea Allan's *The Hookers of Kew* (1967) and W.B. Turrill's *Joseph Dalton Hooker* (Nelson, 1964) provide excellent background, but the main sources are J.D. Hooker's own amazing *Himalayan Journals: Notes of a Naturalist in Bengal, the Sikkim and Nepal Himalayas, the Khasia Mountains &c.*, 2 vols (John Murray, 1854) and *Life and Letters of Sir Joseph Dalton Hooker* edited by Leonard Huxley (John Murray, 1918). The exchanges with Darwin are in Darwin's *Correspondence*, vol. 4, *1847–1850.*

CHAPTER 6: WALLACE AND THE KING BIRD OF PARADISE

A.R. Wallace's *The Malay Archipelago* was first published by Macmillan in 1869. I have used the Oxford University Press edition first published in 1986 with an introduction by John Bastin. Wallace's journal is in the library of the Linnean Society of London. In addition to the books on Wallace mentioned in relation to Chapter 3, there is an important study by J.L. Brooks, *Just Before the Origin: Alfred Russel Wallace's Theory of Evolution* (Columbia University Press, 1984). Arnold C. Brackman in *A Delicate Arrangement* (Times Books, 1980) champions Wallace at the expense of Darwin. Five of Wallace's letters to Samuel Stevens, his agent, are in the Cambridge University Library (Add. 7339/232–36). For the Rajah of Sarawak, see Spencer St John, *The Life of Sir James Brooke* (Blackwood, 1879). For the exchanges with Darwin, see *Darwin* by Adrian Desmond and James Moore, and Darwin's *Correspondence*, vol. 7, *1858–1859.* For Wallace's other papers and articles, I have drawn on *Alfred Russel Wallace: an Anthology of the Shorter Writings* (Oxford University Press, 1991).

CHAPTER 7: THE SAVAGE APE

For the background to *Frankenstein*, I have used the brilliant introduction by Marilyn Butler to the World's Classics edition (Oxford University Press, 1994). For an account of the London Zoo, there is Wilfrid Blunt's entertaining *The Ark in the Park: the Zoo in the Nineteenth Century* (Hamish Hamilton, 1976). For Charles Waterton, see Julia Blackburn, *Charles Waterton: Traveller and Conservationist* (The Bodley Head, 1989), and Waterton's *Essays on Natural History*, 3 vols (Longman's, 1838). Paul du Chaillu's *Explorations and Adventures in Equatorial Africa* was published by John Murray in 1861; I have used the second edition which was issued, with a few corrections and an explanation about the dates, later the same year. Richard Garner, experimenter and naturalist, wrote *The Speech of Monkeys* (1892) and *Gorillas and Chimpanzees* (1896). T.H. Huxley's essays *Man's Place in Nature* were published by Macmillan in 1863. See also Adrian Desmond's *Huxley* and, more generally, the same author's *The Ape's Reflexion* (Blond & Briggs, 1979).

CHAPTER 8: NATURAL PERSPECTIVES

There are several excellent books on lady travellers, notably Dorothy Middleton, *Victorian Lady Travellers* (Routledge, 1965), and Dea Birkett, *Spinsters Abroad: Victorian Lady Travellers* (Basil Blackwell, 1988). Dea Birkett also wrote *Mary Kingsley: Imperial Adventuress* (Macmillan, 1992); another account is Caroline Alexander's *One Dry Season: in the Footsteps of Mary Kingsley* (Bloomsbury, 1989). For Alexandra Tinne, I referred to Penelope Gladstone, *Travels of Alexine: Alexine Tinne, 1835–1869* (John Murray, 1970), and to James Wellard, *The Great Sahara*. Mary Kingsley's own works are *Travels in West Africa* (Macmillan, 1897) and *West African Studies* (1899); there is an abridged text with an introduction by Elspeth Huxley in Everyman's Library. Marianne North's *Recollections of a Happy Life* were published by Macmillan in two volumes in 1892, edited by her sister Mrs John Addington Symonds, followed by *Some Further Recollections of a Happy Life: Selected from the Journals of Marianne North chiefly between the years 1859 and 1869* (from the same team) in 1893. Books on Marianne North include *A Vision of Eden: The Life and Work of Marianne North*, with a biographical note by Brenda Moon (Royal Botanic Gardens, Kew/Webb & Bower, 1980), and Laura Ponsonby, *Marianne North at Kew Gardens* (Royal Botanic Gardens, Kew/Webb & Bower, 1990). The Ranee Margaret of Sarawak's recollections of Marianne North appeared in *Good Morning and Good Night* (Constable, 1934).

CHAPTER 9: A NEW MYTHOLOGY

Charles Kingsley's *The Water Babies* was first published in serial form in 1862. I have quoted from the 1908 edition with illustrations by Margaret Tarrant, and the World's Classics edition

(Oxford University Press, 1995) with the valuable introduction by Brian Alderson. For Maria Edgeworth's *Ennui*, I have drawn on the introduction by Marilyn Butler to the Penguin Classics edition of 1992. For Mrs Gaskell's *Wives and Daughters*, I referred to Angus Easson's introduction to the World's Classics edition, 1987, and to Jenny Uglow's biography *Elizabeth Gaskell: a Habit of Stories* (Chatto & Windus, 1993). For John Buchan, see Janet Adam Smith's biography (Rupert Hart-Davis, 1965). For Conrad, I used the 1990 World's Classics edition with an introduction by Cedric Watts. More generally, I am indebted to Edward Said, especially to *Culture and Imperialism*. For Stevenson, I have quoted from the Penguin Classics edition of *Dr Jekyll and Mr Hyde and Other Stories* with an introduction by Jenni Calder (1979) for 'The Beach at Falesá', from *Robert Louis Stevenson and 'The Beach at Falesá': A Study in Victorian Publishing* by Barry Menikoff (Stanford University Press, 1984), and from the edition of *In the South Seas* introduced by Kaori O'Connor (K.P.I., 1990).

CHAPTER 10: THROUGH THE LOOKING-GLASS

For the cavortings on the beach in South Island, see my biography *Samuel Butler* (The Hogarth Press, 1990). Sir Harry H. Johnston's *A History of the Colonization of Africa by Alien Races* was published by Cambridge University Press in 1899 (sixpence at David's bookstall in 1963). For Thoreau, I have quoted from the edition of *Walden and Civil Disobedience* edited by Sherman Paul (Houghton Mifflin, 1960) (bought in Concord). Simon Schama's *Landscape and Memory* (HarperCollins, 1995) is illuminating about Thoreau, and much else, and his *Citizens* (Viking, 1989) about Rousseau and Ermenonville.

Index